DYNAMICS
OF
AGRICULTURAL DEVELOPMENT

VOLUME-I : LAND REFORMS, GROWTH AND EQUITY

Vol. I	:	Land Reforms, Growth and Equity
Vol. II	:	Technological Changes and Sustainable Development
Vol. III	:	Policy Planning and Liberalisation

About the Authors

Dr. K.S. Dhindsa is a Professor of Economics at the Punjab School of Economics, Guru Nanak Dev University, Amritsar (India). Dr. Dhindsa has a brilliant career, who after graduating from Punjab University, completed the post-graduate diploma in Economic Development from Glasgow University. Thereafter, he obtained his Master Degree from Leeds University (England). Then he went to London University where he successfully completed his Ph.D. degree. He has published a book titled "India's Export Performance—Some Policy Implications". He has published a fairly large number of research papers in National and International journals. For his post-doctorate he went to the School of Oriental and African Studies, University of London, during 1989-91. He was the Editor of *Indian Journal of Quantitative Economics*—A Journal of International Development, during 1994-96. He was the Dean of Faculty of Economics and Business during 1992-93 and was Chairman of Punjab School of Economics, during 1994-96.

Anju Sharma is a Lecturer in Economics in the Post-Graduate Department of Economics, Khalsa College, Amritsar (Punjab). She obtained her M.Phil and Ph.D. degree from Punjab School of Economics, Guru Nanak Dev University, Amritsar. She has carried out a variety of research in the field of Economics. She is the author of a book entitled. *Dynamics of Cropping Pattern : A Regional Study of Punjab.*
A dozen of her research papers have already been published in reputed Journals.

DYNAMICS OF AGRICULTURAL DEVELOPMENT

VOLUME-I

LAND REFORMS, GROWTH AND EQUITY

K.S. DHINDSA
ANJU SHARMA

CONCEPT PUBLISHING COMPANY, NEW DELHI-110059

ISBN : 81-7022-835-2 (Vol. I)
ISBN : 81-7022-834-4 (Set)

First Published 2001

Published and Printed by
Ashok Kumar Mittal
Concept Publishing Company
A/15-16, Commercial Block, Mohan Garden,
New Delhi-110059 (INDIA)
Phone : 5648039, 5649024
Fax : 091-11-5648053
Email : publishing@conceptpub.com

FOREWORD

Development of the agriculture sector in India has traversed a period of five decades since the country achieved Independence in 1947. During this period the very response system of governance has undergone a discernable change from governing an important colony to managing of an independent society that had an entity as a nation state, which was riddled with some formidable problems on economic, social, culture and political fronts. It was under this changed response scenario that agriculture sector assumed prime importance in the development process of the Indian economy so that (1) food and fiber needs of the society were met on a sustained basis, and (2) more than 80 per cent of the population living in rural areas who were wholly or partially dependent for their wherewithal on agriculture sector were provided with gainful employment opportunities. It is of course a different matter, the deriving inspiration from socialistic system of the then USSR, industrial sector was encouraged as a future in the development process and with the public sector achieving commanding heights, distributional social justice was aimed at for the populace on economic front. Yet, this approach and overt emphasis on industrial sector, specially the obssession with big industries, was not supposed to dilute the prominency of the agriculture sector; may be out of sheer compulsions of feeding the large and fast growing population of the country.

Thus, on one hand serious effort was made and are as yet afoot to tone up the agricultural sector for higher growth through strengthening the agricultural research system aimed at developing appropriate production technology, farm level extension of the new technology and agricultural education as well as institutional and policy support and on the other hand the agriculture sector was systematically squeezed to yield marketed surpluses, specially of foodgrain to feed the urban population through public distribution system. Thus, agriculture sector acted as a lead sector that yielded capital surpluses and wage good following a classic model of economic growth and development, though sluggish in pace, as postulated for agriculture sector dominated developing societies, this approach manifested through adoption of the process of investment planning and system of five year development plans.

The priority accorded to the agriculture sector was visible in the First Five Year Plan of 1951-52 through 1955-56. The explicit-emphasis, however, got diluted in the Second plan. Even the Third Five Year Plan did not absorb the reality of the situation and agriculture sector did not get adequate allocations. This happened because of good production response of the agriculture sector in the First Five Year Plan. By the time the mistake was realized it was too late and the country had virtually become a 'basket case' by the mid-sixties.

This impelled the policy-makers to accord priority to the agriculture sector on technological improvement through expanding and strengthening the national agricultural research system, including the central and state level agricultural universities and research institutes. The research effort was complemented with reorganisation and toning up of extension network and also with reorganization of agricultural education. In order to boost sustained growth in agricultural production, supportive institutional and policy framework was put in position through a system of minimum support prices and procurement of agricultural produce by the government. The creation of Food Corporation of India (FCI) and Public Distribution System (PDS) was a land mark in the history of Independent India.

The basic purpose of the system was three fold :

(i) Provision of price certainty to the agricultural producer at reasonable (remunerative) level to save him from vagaries of buyer's market in which stands helpless spectator in the post harvest period, when the buyers so often pooled to bring down the market prices to the disadvantage of the producer-seller. With the introduction of the system of support/procure-ment prices, which by and by covered 22 commodities, the government stand obligated to lift whatever quantities the farmers offer for sale to the government at the announced prices. For this purpose Food Corporation of India acted and any loss was to be covered by the Central Government and profit was to be turned over to the government. The FCI acted direct as well as through the state agencies as their agents.

(ii) Supply of foodgrains, sugar, oil and raw material to the urban population at a reasonable prices through the Public distribution system fed through the procurement of foodgrains, specially wheat, rice and sugar appropriated through sugar levies on sugar mills. Raw materials like jute and cotton were procured from the market at support prices and also quite often at market prices through jute corporation of India (JCI) and cotton corporation of India (CCI) respectively for consumption of the public enterprises.

(iii) Creation of buffer stocks for food security because Indian sensetivities were deeply hurt in mid-sixties when it had become difficult for the government to buy foodgrain from the international market. In order to balance these three, often time conflicting, objectives, Agricultural Prices Commission (APC) was created, which was later renamed as Commission for Agricultural Costs and Prices (CACP), with expanded scope.

Initially the system worked well. In fact during first two years the FCI registered net profit. Yet, soon the ruling party's approach to appease the urban consumers resulted in subsidized issue prices at the PDS and foodgrain as well other commodities were supplied to all the consumers holding ration cards without targeting the distribution for the poor. As a consequence the budget provisions for food subsidies started escalating. Since the government opted not to increase issue prices for the consumers out of political reasons the brunt was borne by the farm sector. Government hesitated to increase support/procurement prices and turned the terms of trade against the farm sector. With a view to maximizing procurement of foodgrain at low prices (support/procurement prices) indirect approach was adopted which eliminated the private trader from the grain-markets. Drastically reducing the limit on food credit (selective credit), putting restrictions on holding of stocks, denying railways wagons to the traders and putting restriction on inter-state or inter-zone movement of grains the market prices were artificially brought down to the level of support/procurement prices. This process continued throughout during the past three decades with exception for only a few years intermitantly when farmers reaped bumper crops, stores/godowns of the government were full and consumer demand and liftings from PDS were low. These years too the farmers suffered on account of back of market clearance for their produce. Thus on one side the production was encouraged as a policy and on the other produce was procured for consumers on low prices to keep the urban voters happy. Thus a great deal of politics, rather cheap politics got mixed with the economics of the policy, which did not put the agriculture sector on an optimal growth path, put a drain on the budgetary resources of the government and created unwanted and unwarranted expectations in the minds of the urban consumers, which promoted pressure on the political system to keep consumption subsidized without any targeting for the poor.

As a consequence of these policy distortions, the agriculture sector remains net taxed or under negative protection according to some scholars. Although this view is widely accepted, yet there are other economists who believe that agriculture sector be brought under the income tax net. There is some logic in this argument too, because there are quite a large number of big farmer who earn huge incomes from farming, yet the sector as a whole is not capable of yielding any income tax, specially when collection costs are kept in view.

There are several other problems the agriculture sector faced as it progressed through last five decades and bore the impact of policy prescriptions at various stages of development. At this stage the farm sector is on crossroads, where production is not picking up as fast as is the need of the nation; weather as yet remains a very important factor and minor deficiencies in rain put the growth of the sector of the track, research and technology development has gone sluggish, policy has gone into a *rut*, farmers are coming under economic squeeze, consumers are becoming more demanding, population growth and improvements in money incomes are increasing the demand for food and fibers, government exchequer are coming under financial strain and above all due to progressive liberaliztion and globalisation of the agriculture market, the agriculture sector is going to face global competition. These are a few of the challenges that are staring at the face of the policy makers and market players in this sector.

It is against this background that the subject matter of this book assumes topical importance. The coverage of the subjects and issues like institution changes in historical perspective, reflections on input out put relationships and their impact on agricultural growth; impact of price support and market intervention, relationships between technology and growth as well as equity, implications and possibilities of agricultural taxation and changing terms of trade, impact of agricultural growth on labour absorption, specially with the adoption of improved technology, probable impact of liberalization and globalization on the agriculture sector as well as general review of macro and micro level agricultural policies followed through this period have a great relevance today if this sector is to be put on an optimum growth path in order to meet the fast growing demand for food and fibre in the national market.

This book is a good job done in this respect; which should prove to be a good material for students and researchers engaged in the field of agricultural economics and specially for those specializing in agricultural development policy.

S.S. JOHL
Prof. of Eminence in Economics (Retd.) and
former Vice Chancellor
Punjabi University, Patiala

PREFACE

There exists at present a considerable body of literature on Indian agriculture in different research publications. Some of this material is accessible in a few edited volumes, but none of them covers all the major, constantly swelling aspects of this subject. So the necessity for editing a comprehensive book on Indian agricultural development was being felt since long. The present book "Dimensions of Agricultural Development" is a humble attempt towards this end. In this book an attempt has been made to include good quality articles of renowned economists published in various reputed Journals. These studies, although related to India's agricultural development can be read with keen interest by various economists and policy-makers who are interested in agricultural development of other third world countries.

While selecting various articles for this volume, syllabi of agricultural economics for post-graduate classes, prepared recently by various universities of India was also kept in view. This book will be of immense use to the students, teachers, researchers and policy makers in India as well as abroad.

In order to make a detailed study of various dimensions of agricultural development in India, thirty three articles have been included in three different volumes. Articles included in these volumes cover both micro and macro aspects pertaining to Indian agriculture, examine the issues and problems empirically and also provide careful presentation of most of the relevant facts of India's agricultural development.

The first volume of the book deals with the issues relating to land reforms, growth and equity. The second volume comprises of various articles relating to technological development in agriculture and sustainability of agricultural development. Third volume deals with the policy and planning concerning agricultural development and the impact of recently concluded GATT round on Indian agriculture.

The main introduction of the book has been given in the first volume. Most of the articles included in these volumes are reprinted articles and some of them are specially invited papers. Details of the original source of each articles has been given in the notes at the end of the chapter.

In an edited book, the co-operation of all those people is very much required whose work we wish to include. We must express our thanks to the contributors for giving us the permission to reproduce their research papers in this book. We are also grateful to the publishers of the Journals who have permitted us to reproduce these papers from their publications.

Before finalising the contents of the book, we had discussions with professors Yoginder K. Alagh, G.S. Bhalla, G.K. Chadha, Krishna Bhardwaj, S.S. Johl, K.L. Krishna, V .M. Rao, Suresh S. Tendulkar and Autar Singh Dhesi. We gratefully acknowledge their valuable suggestions. Thanks are also due to the staff members of Rattan Tata Library, Delhi, School of Economics, Institute of Economic Growth Library (Delhi), National Council of Applied Economic Research (Delhi), Jawaharlal Nehru University Library (Delhi), and Punjab Agricultural University Library (Ludhiana) for providing us facilities to use the relevant research material.

Last but not the least, we are profoundly grateful to the publisher M/s. Concept Publishing Company, of their whole-hearted co-operation in bringing out this book in record time.

K.S. DHINDSA
ANJU SHARMA

CONTENTS

LIST OF CONTRIBUTORS

Anju Sharma
Lecturer, Department of Economics, Khalsa College, Amritsar.

G.S. Bhalla
Professor Emeritus, Centre for the Study of Regional Development, Jawahar Lal Nehru University, New Delhi

Sheila Bhalla
Professor, Centre for Economic Studies and Planning, Jawahar Lal Nehru University, New Delhi

G.K. Chadha
Professor, Centre for the Study of Regional Development, Jawahar Lal Nehru University, New Delhi.

M.L. Dantwala
Professor of Agricultural Economics, University of Mumbai, Mumbai

K.S. Dhindsa
Professor, Punjab School of Economics, G.N.D. University, Amritsar

S.S. Gill
Professor, Department of Economics, Panjabi University, Patiala.

A.S. Kahlon
Professor (Retd.) Department of Economics & Sociology, Panjab Agricultural University, Ludhiana

G. Parthasarthy
Director, Institute of Development and Planning Studies, Muvalavanipalam, Visakhapatnam, Andhra Pradesh

V.M. Rao
Former Member, Agricultural Costs and Prices Commission, New Delhi

H.S. Shergill
Professor, Department of Economics, Panjab University, Chandigarh

D.S. Tyagi
Former Chairman, Agricultural Costs and Prices Commission, New Delhi

A. Vaidyanathan
Professor Emeritus, Madras Institute of Development Studies, Madras

LIST OF TABLES

INTRODUCTION

K.S. DHINDSA & ANJU SHARMA

Agriculture forms the backbone of Indian economy and its dominating position in the national economy can be adjudged from the fact that it contributes a major share of national income and provides livelihood to over seventy per cent of the population. This sector fulfils the food requirements of rapidly growing population and supplies manpower and raw material to India's leading industries.

However, on the eve of independence, agriculture in India was in a hopeless and miserable condition. Agricultural sector was backward and qualitatively traditional in nature. The productivity per hectare and per labour was very low. Country was not self-sufficient in the production of foodgrains but had to depend on imports of foodgrains. The farmers were living in deplorable conditions. Except in certain areas they were dependent upon the vagaries of monsoons and were generally in heavy debt to the village money lenders. Besides, India had suffered a lot because of the partition in 1947, as it was given more people to feed and relatively less land to support them.

Keeping in view the need for rapid development of agricultural sector, policy-makers chalked out the development strategy to raise the production and productivity level of various agricultural crops. The strategy for agricultural development was designed to meet the objective of self-sufficiency in foodgrains. Apart from increase in production, attempt was made to generate additional employment opportunities in agricultural sector for increasing the income of the poorer sections in the rural areas.

To realise the above mentioned objectives, priority was given to the development of rural infrastructure and irrigation projects in various five year plans. In case of both, centre and state plans a large part of public investment was allocated to finance various irrigation projects. In order to encourage private investment in irrigation, like installing diesel and electric pump sets, the government provided cheap institutional credit to the farmers. Power tariff and canal irrigation charges were also kept at low levels (through subsidies) than the cost of their operation. Keeping in mind the role of scientific research and technology, substantial amount of funds were allocated to the Indian Council for Agricultural Rcsearch (ICAR). Many agricultural universities were set-up in different states of India to boost agricultural research at the state level.

Indian agriculture remained in a state of stagnation and backwardness because of the existence of feudal land tenure system introduced by British government. The institutional set-up in Indian agriculture involved the exploitation of actively working group in rural areas, so much so that the actual tiller of soil did not have any inducement to bring about improvement in the agriculture. In view of these problems creating impediments in the growth of entire economy, much emphasis was placed upon institution of agricultural reforms in the country. The various land reforms were initiated for the progressive expansion of agricultural production and income through the reorganisation of agriculture to eliminate all forms of exploitation and social injustice. For this purpose, *Zamindari* system was abolished and Land Tenancy Act was passed to give occupancy rights to the tenants. In order to increase the size of agricultural farms, efforts have been made towards the consolidation of segmented agricultural land holdings. For removing excessive concentration of land with a very small group of big landlords, Land Ceiling Act was passed. These ceiling laws provide for the acquisition of surplus land by the government through the payment of compensation to the land owners. The land thus acquired by the government is distributed among the landless tenants.

Earlier the Indian cultivators were dependent entirely upon the indigenous money-lenders for their financial needs. These money lenders used to obtain bonds and promissory notes from their debtors on false pretence and enter in their sums larger than actually lent. They give no receipt to the farmers for repayments and often charge exorbitant interest rates from the poor farmers. In order to make the cultivators free from the clutches of money-landers, government initiated the policy for the development of credit institutions in various parts of the country in the beginning of the planned era. Thus, the government policy of replacing informal sources of credit by institutional credit has played a crucial role in promoting agricultural development. These institutional credit agencies provide loans to the cultivators at concessional rates. Under various poverty eradication programmes, the government initiated various schemes for providing loans to the poor people. As a result of the promotional role of rural credit institutions, the share of institutional credit which was around 11 per cent during the first plan period rose to 60 per cent during eighties.

Macro economic policies, particularly trade, fiscal and monetary policies were also tailored to achieve the objective of agricultural growth. In order to maintain sufficient supply of foodgrains for domestic consumption, trade in agricultural goods has been canalised. Barring a few traditional crops, the imports and exports of agricultural commodities are handled by government agencies. Only the export of those commodities are allowed which are in surplus over the domestic consumption requirements. In case of domestic shortage of certain essential commodities like cereals, edible oils and sugar etc., their imports are made to keep the prices of these essential commodities stable.

Orderly and well managed marketing acts as an important incentive for increased agricultural production and its commercialisation. Unfortunately, earlier the agricultural marketing system in India was unsatisfactory and exploitative. Poor and illeterate farmers had to confront with powerful and well-organised mass of clever intermediaries for the sale of their produce in the market, who exploited and cheated them in a variety of ways. In order to remove the flaws and malpractices in marketing system and to increase commercialisation of various agricultural products, Government of India has initiated a number of steps. The important ones being (a) The establishment of All-India Warehousing Corporation to construct and manage a network of warehouses in towns and mandis and to increase Warehousing facilities in villages, the co-operative societies are given financial and technical facilities, (b) the establishment of regulated markets, (c) provision for grading and standardisation of produce, (d) standardisation of weights and measures, (e) daily broadcasting of market prices of crops on All India Radio, and (f) improvement of transport facilities.

In order to make agriculture a profitable business and encourage the farmers to increase agricultural production, the government has fixed remunerative minimum support prices for major foodgrain commodities. Even since its inception, the commission for agricultural costs and prices, has been announcing minimum support prices, procurement price and issue prices for a number of agricultural commodities. Minimum support prices are the nature of a guarantee to the producers so that in the event of a glut due to bumper crops, prices are not allowed to fall below these announced minimum prices. To ensure this, government purchases a large quantity of foodgrains at the announced prices. Procurement prices are fixed at a higher level as compared to minimum support prices and are meant for the purchase of foodgrains as required by the government for maintaining the Public Distribution System (PDS) and for building up of buffer stocks. The assurance of minimum support prices and its effective implementation made progressive farmers in the green revolution belt to substantially increase the production of major cereals like wheat and rice.

Apart from giving minimum support prices for increasing agricultural production, the Indian planners have kept prices of foodgrains low through subsidies. The government created Food Corporation of India (FCI) for procuring foodgrains at the fixed procurement prices and then distributing the same to the consumers through Public Distribution System (PDS). Food-stocks procured by FCI are used not only for the PDS, but are also used for releasing it in the open market to stabilise domestic prices at the time of shortage. Earlier, the coverage of PDS was only upto urban poor. But now, its coverage has been extended to rural poor as well.

Recognising the importance of agro-based industries in terms of creating value-added and generation of employment opportunities for people in rural areas, several measures were taken to promote the development of agro-processing industries. After the advent of Green Revolution in mid-sixties, a large number of agro-inputs industries producing fertilisers, pesticides and insecticides etc. came into existence. Also a large number of agro-processing industries like wheat flour mills, rice shellers, oil-mills, sugar mills etc. came into existence. Recently, the Indian government has also allowed the entry of multi-national companies in modern agro-processing industries.

Thus, as a result of large infrastructural investment in irrigation roads, railways, rural electrification, research and extension services, the growth rate of agricultural production and productivity has increased significantly. Hence, the implementation of various agricultural policies followed in the post-independence period has changed the India's position from a food-deficient country to marginally self-sufficient in foodgrains.

Despite remarkable developments in the field of agriculture, critics point out certain failures of agricultural development strategy in India. First, it is said that the technological development in India has neglected unirrigated and backward regions and dry land farming. Besides, it has been limited only to a few crops. Agricultural scientists have primarily focussed their attention on evolving technology for irrigated areas. Also, the agricultural development in areas like dairy farming, poultry farming, horticulture and floriculture etc. has been relatively neglected. Even though, there has been substantial growth in agricultural production, this has not been smooth, instead, there have been continuous fluctuations in crop output from year to year. Furthermore, increase in productivity recorded by India under the impact of modern seed-water-fertiliser-pesticide technology is much below the increase in productivity recorded by the other developing countries like China. With the introduction of modern technology to areas with assured irrigation facilities, oilseeds, pulses and coarse cereals etc. have been shifted to inferior land. Hence, these particular crops have not shown any significant increase in their productivity. In addition to this, there is a severe dearth of appropriate technologies suited to small farmers under risky conditions.

Secondly, land reforms have not been implemented properly due to lack of political will and inefficient and corrupt implementing machinery, which has favoured large and medium farmers who exercise much political influence. The land reforms could not improve the conditions of the poor in rural areas on account of the absence of any strong organisation among tenants, small farmers and landless workers. Consolidation of land holdings has been done on less than fifty per cent of the cultivable land. Besides, because of lack of proper implementation of land reforms, land distribution has remained very much inequitable and thus the land base of the bulk of cultivators is too small to generate sufficient income for their family sustenance. Furthermore, though many efforts have been made to regulate tenancy, it still prevails at large in various parts of the country in the concealed form.

Thirdly, agricultural policies have increased intra-regional and intra-personal inequalities. It is due to the fact that new technology comprising high yielding varieties of seeds, fertilisers, and pesticides etc. was devised only for irrigated land areas. Rain-fed areas and agriculturally backward areas lacking assured irrigation facilities have not been able to take the advantage of this technology. As far as intra-personal inequalities are concerned, one important reason for this is the skewed distribution of land holdings. Large and medium farmers with large size of land holdings have gained much more than the small farmers because these poor farmers have not been able to get full benefit from new agricultural technology because of the lack of their adequate access to institutional credit and advisory services.

Fourthly, the benefits of Public Distribution System which were targetted to urban poor only were derived by all the ration-card holders, poor and non-poor. However, only recently, the rural poor have been covered under the network of Public Distribution System. Even the targetted Public Distribution System has been evolved, but still there are many flaws in its institutional and operational set-up. It is also pointed out that this system has failed to serve the poor eastern states of India.

Fifthly, only the developed regions of the country have cornered benefits from credit institutions. Even in the developed regions only influential farmers have gained the bulk of the benefits from formal rural credit institutions.

Sixthly, agriculture sector did not get adequate amount of plan outlay, especially during the Second and Third Five Year Plans. Real public investment in the agricultural sector has also been declining continuously since last two decades.

Seventhly, due to the high protection given to the non-agricultural sector, inter-sectional terms of trade have been unfavourable for the agricultural sector. This has acted as a serious hindrance in the way of agricultural growth.

Eighthly, due to slow and inadequate development of secondary and tertiary sector in India, surplus labour from agriculture sector could not be shifted from this sector to other sectors. This has increased the dependence of labour on agriculture sector.

Lastly, there has been a critical contradiction in the agricultural price policy of the country. On one hand, it has tried to encourage the production and productivity and on the other hand, it kept deliberately the prices of certain agricultural goods at lower levels. Furthermore, the minimum support prices have not been fixed for all the crops. FCI has also become unviable due to the inefficient and corrupt administration, huge size of its operations and the requirements of a large amount of subsidies to support it.

The above discussion clarifies that despite remarkable progress made by Indian agriculture in the past fifty years, Indian agriculture still has to face many challenges in the times to come, namely;

(a) increase in agricultural production at a faster rate so as to meet the food requirements of the rapidly growing population.

(b) establishment of various credit institutions for providing cheap credit facilities for agricultural development in rural areas.

(c) development of suitable technology for rainfed and dry land areas.

(d) absorption of surplus labour in agricultural and allied activities
(e) evaluation of the existing public distribution systems and streamlining it for deserving people.
(f) possibilities of mobilising resources from the agricultural sector through taxation
(g) reviewing the existing price support policy for the benefit of poor farmers.
(h) initiating the process of shifting disguisedly unemployed labour force from agriculture sector to non-agricultural sector.
(i) strict implementation of land reforms to achieve higher agricultural growth rate and the improvement of the lot of a large number of small and marginal farmers.
(j) development of agro-based industries to strengthen the inter-sectoral linkages so as to stimulate the overall growth of the economy.
(k) improvement in the efficiency of agricultural marketing to save the small farmers from exploitation and malpractices of middlemen. Attention to be paid especially on the development of co-operative marketing societies in rural areas.

Apart from the issues discussed above, some critical challenges have been posed before the Indian agriculture after the liberalisation and globalisation of the economy. Particularly, the recently concluded GATT (WTO) agreement and trade policy reforms are bound to have important implications for the Indian agriculture. The key elements of the GATT agreement, which will have impact on Indian agriculture relate to the liberalisation of imports of goods, patenting of seeds, gradual elimination of input subsidies.

It is generally argued that present world prices of many agricultural goods are distorted because of the domestic price support programmes and export subsidies given in case of several agricultural goods. It is pointed out that as a result of the gradual reduction of these subsidies over a period of time, efficient producers both from the developed and developing countries would be able to gain by exporting the products which they earlier could not export because of high tariff and non-tariff barriers imposed by the importing countries.

The impact of liberalisation and GATT agreement on Indian agriculture have been debated by many scholars. Some scholars argue that impact of liberalisation on Indian agriculture would be favourable. They opine that reduction in trade restrictions alongwith the devaluation of overvalued exchange rate will definitely improve terms of trade for agriculture sector and hence boost the production and exports of agricultural goods including processed products. The supporters of liberalisation are of the view that because of the withdrawal of subsidies on temperate products by developed countries, Indian can gain through the exports of temperate products like sugar, cereals, horticulture, dairy and livestocks products. However, for removing domestic shortages of these goods, adequate surplus of these goods over domestic consumption requirements should be created. It is also argued that as the cost of seeds form a small proportion of total cost of cultivation, farmers will be in a position to buy patented seeds if these seeds yield substantially higher output. In this situation the gain achieved due to the improved yield will be sufficient to offset the loss due to the high cost of patented seeds. However, some economists are of the view that liberalisation is going to have negative effects on the agricultural sector in India.

First, it is contended that provisions relating to Trade Related Intellectual Property Rights (TRIPRs) would have an adverse effect on indigenous research and development in agriculture, which in turn may have serious implications for the future growth of agriculture. It is mentioned that prohibition to sell, exchange, propogate and adapt seeds to local conditions would hamper diffusion of new agricultural technology. Secondly, because international prices of foodgrains show great fluctuations than domestic prices, aligning domestic prices of foodgrains with world prices would jeopardise India's food security system. Thirdly, raising the prices of agricultural goods to the level of world prices, would increase the prices of foodgrains which in turn would badly hurt the rural and urban poor. Lastly, it can be said that big farmers and regions with better infrastructural facilities will be the main beneficiaries from GATT and this will further aggravate the inter-personal and inter-regional inequalities.

To conclude, it can be said that Indian agriculture can benefit from liberalisation and globalisation if the government allocate large resources for public investment in rural infrastructure particularly irrigation, rural electrification research and extension. Thus fiscal deficit under the new economic policy, should be reduced by cutting unproductive public expenditure rather than by cutting essential public expenditure incurred on rural infrastructural investment. The government can curtail input subsidies to a minimum level by withdrawing subsidies in a phased manner from the medium and large farmers. But this requires strong political will. Trade subsidy presently given to FCI should be curtailed and for bringing efficiency in the working of FCI, other private/public agencies should be assigned the task of procuring and distribution of essential commodities. The distribution of food through PDS at subsidised rates must be targetted to the poor only. Thus the resources saved from the withdrawl of subsidies should be used as grants to states for expanding irrigation. In order to meet the food requirements of the rapidly growing population and thus to avoid an inflationary price spiral, the production of food grains must be increased at a faster rate. This in turn calls for allocating larger funds to agriculture in real terms. For augmenting agricultural production, minimum support prices for major commodities must be continued in order to protect agricultural producers from fluctuations in prices. As a result of WTO agreement and liberalised economic policy there is a great scope for increasing the exports of rice, cotton, dairy, livestock and other agro-processed products. But this can be possible only, if adequate surplus over and above the domestic requirements is created. For this, productivity and output of these commodities will have to be increased at a faster rate. Indian agriculture has to become cost-effective and internationally competitive. Because of the relatively larger fluctuations in the international prices of agricultural commodities and possibility of shortage of foreign exchange, the policy of keeping food stocks need to be continued for preventing sudden jump in food prices and thus avoiding inflationary price spiral.

In order to make an in depth analysis of major aspects of agricultural development in India, the book has been divided into three different volumes and thirty three articles have been included in the volumes.

India being a large country with pronounced regional variations in terms of soil type, rainfall availability and potential of irrigation water, size of holdings etc., the importance of studies on regional problems of Indian agriculture can hardly be exaggerated. Keeping this fact in mind, some useful studies on regional issues pertaining to Indian agriculture have also been incorporated in different volumes.

The first volume of this book deals with issues relating to land reform, growth and equity and labour absorption. This volume is comprised of 10 articles which have been further grouped under three sections.

SECTION I

Articles included in Section I highlight the impact of institutional changes on agricultural development in terms of various land reforms initiated by the Indian government after independence. These papers cover various aspects of land reforms namely emergence of informal tenancy due to lack of proper land records, land ceilings, reasons for success of land reforms in states like Kerala and West Bengal and suggestions for the proper implementation of these reforms in various states of the country.

In his paper, "Land Reform Experience-Perspective for Strategy and Programmes", V.M. Rao discusses land reform experiences in different states of India and suggests some priority tasks in the next round of land reforms. He points out that one third of the land in India is under concealed tenancies. Informal or concealed tenancies — is a euphemism for landlord-tenant relationship-remaining entirely out of reach of tenancy legislation for security, fair rents and transfer of land to actual tillers. The author also stresses that consolidation of land holdings definitely helped in increasing agricultural production particularly through investment in the development of irrigation sources by private individuals. Since only 40 per cent of the total cultivated land has been consolidated in India, so the government must initiate measures to prevent the fragmentation of holdings. In Bihar state, there is an urgent need for conferring tenancy rights to share croppers. The land reforms in Punjab did not impose an effective ceilings on the size of holdings, and these did not succeed in preventing the large owners from resuming land for self-cultivation. The process would eventually come to an end with the small owners voluntarily selling their land holding to the large tenant.

Rao mentions the following priority tasks for the next three rounds of land reforms : (i) consolidation of holdings to be completed in all parts of India, (ii) good land records are to be kept (iii) all lease transactions must get properly recorded and their terms and conditions should be made fairer to weaker party (iv) land reforms to be supplemented by research, extension and credit services etc.

In his paper entitled, "Lease Market, Poverty Alleviation and Policy Options", Parthasarthy points out that despite significant leasing in by large owners, the dominating and persisting trend in this group is towards leasing out. Author alsc mentions about the growing trends of capitalistic tenancy in which medium and large farmers lease in land. Capitalistic tenancy (reverse leasing) restrict the access of marginal owners and the landless farmers to the lease market. The author stresses that the bulk of small tenancy is concentrated in the states with relatively high incidence of poverty. For example in Bihar, the inverse relationship between lease-in households and the size of holding has been found. Moreover, in areas of high concentration of tenancy, share cropping arrangements, under which all costs continue to be borne by tenants, is on the rise. Thus an effectively implemented tenancy legislation has not only the potential for poverty alleviation but could also contribute to growth. According to author there are two pre-requisites for the success of land reforms (a) political will for successful implementation of the reforms and (b) successful mobilisation of the beneficiaries in favour of the reforms.

Sucha Singh Gill in his paper has discussed about the changing land relations in Punjab and their implications for land reforms. He emphasises that land relations are very much important and have implications for production, productivity and distribution of income and wealth in agriculture sector. While commenting on land reforms in Punjab, the author points out that with enactment of various acts, occupancy tenants have got the ownership rights. According to tenancy laws, a tenant could not be ejected from a holding upto five acres. However, the rate of compensation to acquire ownership of land was too high in Punjab and was beyond the capacity of poor tenants. Also, the implementing machinery has been biased in favour of big land owners and the influential persons have sabotaged 'Ceilings on Land Holding Act' by acquiring land in the names of relatives, obliging tenants or landless Harijans. Laws were also made for compulsory consolidation of holdings to solve the problem of fragmentation.

The author pinpoints that the decade of seventies has brought remarkable changes in the land relations in Punjab. Land is shifting from small holders to large holders—a capitalistic development tendency of concentration of productive resources in the hands of a few producers of large size. The evidence suggests that the shift of land from small and marginal farmers to medium and large farmers is on account of leasing out of land. It reveals that most of the leasing out is done by small owners and leasing in by the big operators. Gill observes that in order to protect the weaker farmers, some modifications in the present tenancy act should be made. First, the operation of Tenancy Act for the protection of tenants against eviction must be restricted for small cultivators. Secondly, for usefulness of ceiling laws and land redistribution among the poor, there is a need to take steps to check tendency of small farmers going out of operation. For this purpose, small producers should be protected through various measures of the government i.e. by providing assistance and facilities serving exclusively the needs of such farmers. The author stresses that the only way to check liquidation of poor/small cultivators is to introduce land reforms on Soviet and Chinese pattern. It is some sort of collectivisation which can protect the poor cultivators.

In a commercialised agrarian community, economic competition among peasant families occurs mainly in the land market. The question arises is that why some peasant families are able to accumulate wealth and buy land while many others fall in chronic 'debt' and sell their land holdings. H.S. Shergill has addressed this particular issue in his paper and identified the factors responsible for the 'growing' and 'decaying' peasant farms in Punjab. His findings reveal that the change in the owned area of peasant families through land market transactions is positively influenced by per capita income, number of adult males in the family, financial position at birth, money lending and early adoption of new technology; and is negatively influenced by addiction of family males to wine/opium and larger size of owned area at birth. In a peasant cultural environment, the pressure to acquire more land and retain the already owned land is likely to be inversely related with per male member area.

SECTION II

Included in Section II are the articles discussing important issues like growth and equity, impact of technological changes on productivity and suitable new technology for the development of dry land agriculture.

M.L. Dantwala in his paper points out that though overall performance of Indian agriculture is quite satisfactory, even then there are considerable inter-regional inequalities in the growth. The bulk of foodgrain production has come from a relatively small area-the so-called Green Revolution belt. He stresses that for removing inequality and poverty, inequality in asset ownership both in rural and urban sector has to be reduced. Author also argues that because agro-climatic environment varies vastly from region to region and the irrigation potential is not uniformly distributed, therefore there is no economic sense to seek product uniformity for all regions of the country. If certain regions have comparative advantage in the production of foodgrains but their potential is untapped, then efforts must be made to increase productivity of only foodgrains in these regions.

G.S.Bhalla and G.K.Chadha have evaluated the impact of Green Revolution on income distribution among cultivating households in

Punjab. The authors point out that Green Revolution in Punjab has brought overall prosperity to its peasantry. However, the gains of new seed-irrigation-fertiliser-pesticide technology have been distributed, more or less, in proportion of the initial land holding position. Since the land holdings are distributed in a highly skewed manner, the distribution of gains has also been quite inequitable. They argue that since low income of marginal and small farmers is due to small land base, these farmers should be involved in other allied and non-agricultural activities. It is also stressed that one of the main reasons for rural poverty is overpopulation in agriculture combined with inequality in land distribution. The authors emphasise that real solution for the eradication of rural and urban poverty lies in the rapid industrialisation and diversification of the economy, which in turn requires capital formation at a high rate.

For sustaining agricultural growth over the coming years and to develop stagnant regions, breakthrough in dry land agriculture is a must. V.M.Rao's paper is exclusively related to the development of dry land agriculture. He argues that in case of dry land agriculture, increases in land productivity can be achieved by conserving soil and moisture and by choosing crop/crop varieties having the potential for higher productivity under dry land conditions. It is suggested that in addition to strong research programme for technological changes, dry land agriculture needs effective implementation of minimum support price policy and credit delivery system. It is felt that tank irrigation, which in the past was maintained by village efforts, is now dying at an alarming rate. In order to improve irrigation facilities in these dry regions attention should be paid towards the development of watersheds which require an amount of Rs.2000 per hectare whereas land development in command areas require a huge amount of about Rs.12000-15000 per hectare.

In their paper on Inter-sectoral Terms of Trade, A.S. Kahlon and D.S. Tyagi point out that many past studies on terms of trade have either used wrong methodology and/or inappropriate data. According to them, for obtaining accurate estimate of terms of trade, the following conditions must be fulfilled, namely (a) appropriate choice of base year, (b) selected price indicators must reflect the timely movements of prices both for exports and imports, implying that consumers price index rather than wholesale prices should be used, (c) use of current/base period quantities as weights for constructing price indices.

The authors argue that the use of wholesale price index for

determining terms of trade is not proper because the weights assigned to different commodities in the construction of wholesale price index are based on total volume or transactions and do not reflect the quantity sold by one sector to the other. For constructing price indices, the authors have made use of farm harvest prices of various agricultural commodities sold by the farmers. Using Laspyre's (current year weights) formula and weights derived from NSS and CSO data separately, the authors have calculated the net barter terms of trade for all commodities and sub-groups. Their results clearly show that the terms of trade have been unfavourable to agriculture during the period under study. They also emphasise that the unfavourable terms of trade can adversely affect the level of technological inputs used and hence the capital formation and the level of output in the agricultural sector.

SECTION III

The articles in Section III relate to an important issue regarding the absorption of labour force in Indian agriculture. These papers also provide us the estimates of employment with respect to total agricultural output as well as crop-wise elasticities of employment with respect to yield etc.

Shiela Bhalla in her paper entitled, "Trends in Employment in Indian Agriculture, Land and Asset Distribution; points out that despite shift in cropping pattern towards labour intensive crops and increase in gross cropped area, total labour absorption has been either stagnant or may have fallen in absolute terms in several technologically dynamic states. After mid-seventies, a declining trend in labour use was observed and this could be the result of widespread adoption of mechanisation, without which in time sowing of second crop was not possible. Thus there is a conflict between goals of maximising output on one hand and the employment on the other. The results of her study reveal that in all states of India; sugarcane, paddy, cotton and jute have higher labour requirements per hectare. As long as the spread of labour saving technology is associated with major changes in cropping pattern in favour of relatively labour intensive crops together with increase in gross cropped area, the total employment effect with respect to entire mix of crops or even with respect to particular crops may be positive.

According to Bhalla, for absorbing surplus labour of the agricultural sector, several measures should be taken. First, since per hectare growth in employment is negative in many states, the increase in the present level of employment in crop husbandry sector can be possible through the increase in gross cropped area and change in cropping pattern to more labour intensive crops and this in turn would depend on the pace of development in private and public irrigation projects Secondly, the policy makers should pay adequate attention to the development of agro-based and other industries for absorbing surplus labour force of the agricultural sector.

A. Vaidyanathan's paper on labour use in rural India, examines inter-state and inter-district variations in total labour input into agriculture per unit of area and the reasons for these variations. Cross-sectional data show that states/districts, where the output per unit of gross cropped area is high, tend to use a larger input of human labour per unit of gross cropped area. A significant positive relationship between labour intensity and total rainfall has also been found. The author points out that technological changes alongwith differences in agro-climatic environment are significant factors influencing productivity of land and hence the intensity of labour use. The author also discusses the factors influencing rural non-agricultural employment. The ratio of non-agricultural to total employment in rural areas (in terms of man-days) has been found to vary from 12 per cent in Madhya Pradesh to 30 per cent in West Bengal.

SECTION I

LAND REFORMS

1

LAND REFORM EXPERIENCE

Perspective for Strategy and Programmes

V. M. RAO

I. INTRODUCTION

The theme of land reforms, which figured prominently in the first decade after independence, receded into the background in the latter decades. When the green revolution arrived in the late 60s economists and the policy-makers were quick to persuade themselves that the new technologies held the key to the problems of slow growth as well as to those of socio-economic inequities in rural society. But, after two decades of green revolution, this belief has turned out to be an illusion, in the wake of limitations of green revolution observed in the field, viz., it has remained confined to a few pockets, small number of crops and selected farming groups. The setting up of National Commission on Rural Labour in 1987 is a significant landmark in the changing policy perspective at the highest levels on the problems of rural labour and similar categories of rural poor. A perusal of the themes taken up by the study groups assisting the NCRL would show that issues of institutional and structural reforms in Indian society are coming to the fore in the development strategies for the poor during the 1990s and beyond. The role that land reforms can play in the reorientation of the development strategy is the theme taken up in this paper.

The revival of interest in land reforms has also a second source. The last two decades have witnessed considerable experimentation with the strategy of 'direct attack' on poverty consisting of Minimum Needs, Employment Generation and IRD programmes. The direct attack on poverty was conceived as a complementary approach to stepped-up growth. The two together were expected to help the process of trickle down to gain in speed, spread and thrust. However, the outcome of direct attack strategy has been far too modest as an approach to poverty eradication in growth in the 80s along with substantial expansion in agricultural output. In particular, the core groups of poor like landless labourers are seen to have obtained minimal enduring benefits from both growth and anti-poverty programmes. There is now increasing awareness among social scientists that attempts to integrate growth with poverty eradication remained largely ineffective owing to the absense of adequate structural reforms and improved access for the rural poor to agricultural and common lands.

The performance so far of various measures for land reforms, as will be seen later in this paper, has ranged from modest to disappointing. Their cumulative impact on the rural poor after more than four decades of voluminous and impressive legislation has been disconcertingly meagre. Is the revived interest in land reforms then no more than a passing political whim? This paper adopts a sharply divergent point of view. Its purpose is to build up a strong positive argument for land reforms to improve the status of rural labourers and the poor. The arguments is based on the premise that the very failures of our development strategy so far—slow growth, modest poverty eradication and negligible structural reform—are indicative of the scope available in the Indian economy for integration of these three major components in the development strategy to benefit from their capacity to mutually reinforce each other. It needs to be remembered that the conditions prevailing now are substantively different from the context in which the first major push for land reforms was planned and undertaken in the immediate post-independence years. The foundations now available for growth, for technological modernisation and for extensive and sustained anti-poverty operations mark a significant change in the socio-economic context. The changed context has the effect of creating a new and pressing need for land reforms as well as of providing new and relevant opportunities for their implementation. It is not our case that the land reforms have now become easy to achieve. Our attempt is only to show that, given the changed and emerging context, determined measures of reforms may reasonably hope for more substantive success than in the past. What is important is that the measures be consistent with the requirements of changing situation.

The plan of the paper is as follows. Section II outlines our perception of the changed context with particular reference to elements in it relevant for planning of land reforms. Section III brings together the glimpses of implementation and impact gathered from the growing mass of information and research literature on this theme. Section IV reviews the recent grassroots level trends in the access of rural poor to the common lands and the problems encountered in widening, improving and institutionalising the access. The last section in the paper, after describing the strategy framework which we consider to be relevant and practicable, identifies and classifies the priority tasks to be taken up so that land reforms get focused on the rural labourers and poor. The paper does not pretend to offer ready-made solutions for the challenging problems of land reforms to improve the status of rural poor. We only hope that the paper would be of some assistance to policy-makers and researchers in undertaking a systematic search for solutions and in working out a policy perspective which is pragmatic and avoids both the extremes of naive optimism and corrosive cynicism.

II. SOCIO-ECONOMIC CONTEXT

Planned structural changes like land reforms need to be conceived keeping in mind the socio-economic context. Feasibility of such changes depends on careful prior thinking to adjust the objectives, strategies and programmes to the socio-economic context, particularly the constraints it imposes on policy-induced modifications in structure. Let us first see some telling symptoms of constraints.

It can be seen from Table 1.1 that by 1985-86 over 76 per cent of total operational holdings belonged to 'marginal' or 'small' size-groups as compared to 70 per cent in 1970-71.

Accelerating marginalisation of holding stands out clearly in the table. The percentage share in total holdings of the 'marginal' holdings increased sharply over 1970-71 to 1985-86 (from 51 to 58 per cent). In contrast, the share of size-groups from 'small' to 'large' decreased over the same period.

By way of a quick assessment of the 'redistributive' potential available in the prevailing structure of operational holdings, one can do the following calculations for the year 1985-86.

Assuming a ceiling of 10 hectares, the 'large' size-group (10 hectares and above) would be the only group having 'surplus' which would add up to a little less than 14 million hectares (1.93 million 'large' holdings with an average size of 17.2 hectares could each yield a 'surplus' of 7.2 hectares). Even if all this 'surplus' were transferred to 56.75 million 'marginal' holdings, the average size of holdings in that group would go up from the prevailing 0.38 hectares to only 0.64 hectares. Clearly, with accelerating marginalisation of holdings, the redistributive potential is fast evaporating (Singh, A. K., 1981).

Growing pressure of population on land (indicated by Table 1.1) in an unequal society would mean harsher and increasingly precarious conditions of living for the rural poor. A recent report on tenancy prepared by the National Sample Survey Organisation finds evidence of substantial and growing extent of 'informal' tenancies. The report observes "the land reforms legislation drove it (tenancy) underground. . . . The estimates of leased in area operated was of the order of 2.2 million hectares according to the Agricultural Census, 1980-81 (which was) only one-fourth of the corresponding estimate of 8.9 million hectares thrown up by the NSS survey of 1981-82. This difference is mainly because the Agricultural Census, being record-based, could not take into account the tenancy, which was contracted orally" (Bhattacharya, M. G. 1989). A recent seminar on 'Land Reforms — A Retrospective and Prospect' organised by he Planning Commission mentions a much higher estimate of a third of land (being) under concealed and informal tenancy (Planning Commission, 1989). 'Informal' tenancy is a euphemism for landlord-tenant relationship remaining entirely out of reach of tenancy legislation for security, fair rents and transfer of land to actual tillers. Its growing extent is a revealing symptom of how an adverse socio-economic context could frustrate reformis' measures.

In the prevailing setting, even growth and technological change do not adequately help the poor. The Economic Advisory Council of Government of India mentions in its latest report that during the 80s, which witnessed aggregate output growth at more than 5 per cent per annum, the growth in employment did not reach the level necessary for absorbing the additions to labour force during this period (Economic Advisory Council 1990). Detailed statistical analyses done by Mishra show that the levels of employment and of wages do not necessarily improve when technological change occurs in agriculture (Mishra, G. P., 1988, 1989).

More pertinent to note, the socio-economic context exerts a strong dampening influence on the political will and administrative efficiency with which programmes for land reform, get formulated and implemented. This dimension of the socio-economic context is reflected well in the following observations which, interestingly, come from an assessment of the land reforms programme done within the government of India itself by the department of rural development, ministry of agriculture (Government of India, 1988).

> "It is widely recognised that the chief reason for poor implementation of land reform has been the lack of political will. It would not be surprising to expect so, if we appreciate the realities of the rural situation and the development of Indian politics".
>
> "The rural rich maintain their position through use of muscle power and manipulation of administrative and judicial processes. The poor state of land records and a weak revenue administration also serve the rural rich who hold land in excess of ceiling despite all the land laws and violate tenancy laws as well. The dominant faction in Indian politics is recruited from rural rich. It is lack of political will to act against this class which ultimately characterises the lack of political will of state governments."
>
> "This ambivalence at the political level is matched by an administrative style in the bureaucracy. A good officer (is one who) accomplishes 'token' implementation. . . . (for example) collectors who would allot land to the rural poor by force or legal harassment. Such a style of functioning, accomplishing 'token' implementation, is efficient on paper, and appreciated by vested interests. . . . the role of police and revenue machinery especially at the lower level, in implementation of land reforms has also been negative."

It would be interesting at this point to make a brief reference to a comprehensive class analysis of land reforms programmes in Karnataka done at the Institute for Social and Economic Change (Thimmaiah, G. and others, 1984). According to this analysis, in the earlier years of land reforms immediately after independence, the dominant group opposing land reforms (mostly covertly) was the group of absentee urban elite owning land in the villages from which they had migrated for good. With the changing distribution of political power and emergence of growth pockets in agriculture, the dominant group now in the middle and large

TABLE 1.1 : Operational Holdings in India Agriculture Census

Category and Size	*Number of Operational Holdings (Million)*				*Area Operated (Million Hectares)*				*Average Size of Holdings*
	1970-71	*1976-77*	*1980-81*	*1985-86*	*1970-71*	*1976-77*	*1980-81*	*1985-86*	*1985-86*
1 Marginal	36.20	44.52	50.12	56.75	14.56	17.51	19.74	21.60	0.38
(below 1 ha)	(51.0)	(54.6)	(56.4)	(58.1)	(9.0)	(10.7)	(12.1)	(13.2)	
2. Small	13.43	14.73	16.07	17.88	19.28	20.90	23.16	25.53	1.43
(1-2 ha)	(18.9)	(18.1)	(18.1)	(18.3)	(11.9)	(12.8)	(14.1)	(15.6)	
3. Semi-Medium (2-4 ha)	10.68	11.67	12.45	13.25	30.00	32.43	34.65	36.58	2.76
	(15.0)	(14.3)	(14.0)	(13.5)	(18.5)	(19.9)	(21.2)	(22.3)	
4. Medium	7.93	8.21	8.07	7.92	48.24	49.63	48.54	47.01	5.94
(4-10 ha)	(11.2)	(10.0)	(9.1)	(8.1)	(29.7)	(30.4)	(29.6)	(28.7)	
5. Large (10 ha	2.77	2.44	2.17	1.93	50.06	42.87	37.71	33.19	17.20
and above)	(3.9)	(3.0)	(2.4)	(2.0)	(30.0)	(26.2)	(23.0)	(20.2)	
All categories	71.01	81.57	88.88	97.73	162.14	153.34	163.80	163.91	1.68
(100.0)	(100.0)	(100.0)	(100.0)	(100.0)	(100.0)	(100.0)	(100.0)	(100.0)	

Notes: 1. The figures in brackets are percentage of totals in respective columns.
2. The figures for 1985-86 are provisional

land-owning-cum-cultivating groups within the villages. They pose a political challenge—to the government and to the rural poor—which is far more formidable than the earlier challenge from the absente owners. The dominant groups now emerging on the scene are much more ruthless, have a tighter grip over the policy-making politicians at the highest levels and use extra-legal methods at the ground level more effectively and with greater impunity and disconcerting frequency.

An important implication of the feature of socio-economic context noted above is that the strategy for structural reforms in the rural society—of which land reforms are a major part—which could be effective in the coming decades would have to have two well designed components to improve the socio-economic context to make it more congenial for reforms of rural structures and relationships. First, it would be necessary to strengthen substantially the capacity and willingness of government machinery, personnel and procedures to meet the challenges from the emerging dominant groups. Second, it would be essential to create preconditions which are favourable for the rural poor to acquire adequate countervailing power against the ascendent dominant groups and to build up entrepreneurial capabilities of their own to compete with the rural rich in exploiting new economic opportunities without the crutches of government support and/or aid from external groups.

In the light of the description of the socio-economic context given above, it would be legitimate of ask whether such changes to modify the socio-economic context would be feasible at all now or in the near future. While our response to such doubts is not overly optimistic, we point out a few indicators providing some basis, however modest, for a measure of hope. First, growth in Indian agriculture now confronts barriers which cannot be overcome through primarily technocratic strategies which succeeded so admirably in the green revolution areas. Growth in the green revolution areas is now levelling off. It is also realised that it is being achieved at a high and increasing cost in terms of scarce inputs like irrigation and fertiliser. Growth in dryland agriculture and in the potentially rich eastern parts of India needs a combination of both technocratic and reformist strategies (Rao, V. M., 1986). The point is that, unlike in the immediate post-independence years, the need for land reforms is now far more pressing—since sustained growth itself depends on them—and this may have the effect of strengthening the political will for reforms to make it more assertive.

Second, there have been two favourable fall outs from the extensive implementation of poverty-alleviation programmes—(i) the development personnel have been under fairly severe pressure to establish contact with the poor and to assess their felt needs, and (ii) the rural poor are now in a better position and mood to demand more enduring benefits from the government than *ad hoc* relief during periods of distress. In this sense, PAP can be regarded as being a 'conscientising' influence on both the development personnel and the poor. Third, and most important, there is a subtle difference between the absentee-cum-non-cultivating owners on one hand and the present dominant rural group on the other in the nature of the resistance they put up to land reforms. The former are rent-collectors whose resistance to reforms began to weaken only when they moved out of villages in search of new economic opportunities in towns not depending on land. The latter group, on the other hand, consists of profit-earners who may be expected to respond more readily and positively to new growth opportunities in agriculture and rural sectors. If there are policies to ensure that the new opportunities are adequately labour-absorbing and—through a combination of farm and non-farm enterprises—bring a measure of affluence within the reach of even households with modest land-holdings provided they are enterprising enough to benefit from the new opportunities, the feasibility of effective implementation of ceiling and tenancy legislations might show a marked improvement.

Thus, while the socio-economic context has so far been overwhelmingly adverse to land reforms, some positive changes seem to be in the offing. More important, the policy-maker need not feel as helpless as before in inducing desirable modifications in the context. Despite their modest and mixed achievements, the programmes for growth and for poverty alleviation, themselves, may work towards inducing such modifications (Rao, V. M., 1988). This is the perspective from which we assess the scope for land reforms which are more substantive than in the past and are more effectively focused on the poor.

III. IMPLEMENTATION AND IMPACT

A word about the design of this section. The section is focused on the specific issue of how far the benefits of land reforms reach the rural poor and landless. There is no intention to cover the entire area of land reforms which is wide consisting of innumerable legislations and their revisions and amendments. From the perspective of social science researchers, there are two principal benefits which the rural poor can obtain from land reforms—(i) ownership of land for those who are landless or have meagre holdings (in fact, so meagre that often these tiny owners lease out their land instead of cultivating it themselves), and (ii) security of occupation and fair terms for the poor tillers of land who are tenants. Our attempt is to bring together the glimpses of diverse field situations in this respect by categorising them in an interesting and meaningful way.

It is useful to start with the proposition that, by and large, land reforms have been a failure in benefiting the poor. The official assessment itself is quite pessimistic—"Land Reforms Programme has virtually come to a dead end . . . . there is a strong opinion in the country that land reforms have really not been implemented" (Government of India, 1988). This is a rather extreme view which we do not fully share. The field situations appear to range from areas where the land reforms programmes have been frustrated at the very initial stages of putting the necessary legislations in the statute book to areas where, despite some good and effective implementation and receiving of programme benefits by the poor, there was no enduring impact on the poor and on the system of land relations. Thus, while it would not be wrong to say that the poor did not benefit much from the reforms in concrete terms, the reasons for failures were not uniform in all areas. In particular, it is necessary to distinguish the situations where there was no worthwhile formulation of land reform programme from those where the poor did not get enduring benefits despite fairly serious and effective implementation. In this sense, the Indian setting is rich in diversity. The purpose of this section is to see its implications for the land reforms strategy and priorities.

We find it useful to have the following four-category typology of areas.

(a) 'Semi-feudal' areas like Bihar, where, according to scholars like Pradhan Prasad, the barriers to reforms inherent in the prevailing structure of society are too strong to be overcome by the scale and intensity of efforts for land reforms made so far.
(b) Areas (e.g. Karnataka) with less formidable barriers to land reforms and stronger thrust for reforms as compared to 'semi-feudal' areas but having mixed and, on the whole, modest results of reforms.

(c) Areas (e.g. West Bengal and Kerala) with strong political mobilisation and effective land reforms but where the growth performance has been too weak to complement reforms to bring about enduring structural change.

(d) 'Green revolution' areas (e.g. Punjab) where capitalist farming is firmly entrenched and where reforms are needed mainly to improve wages, work environment and economic conditions of labourers.

We begin by taking note of the reforms situation in Bihar. Abolition of *Zamindari* was the foundation on which substantive improvements in land relations could occur in Bihar. But, this foundation itself remained weak and could not support any worthwhile changes in land relations. The following glimpses come from a perspective paper by Dhar (Dhar, Hiranmay, 1990). "By their chickenery, system of *benamdari*, collusion and their societal power, the ex-zamindars of Bihar succeeded by and large in protecting their class interest. Up to 1983-84, only 1.26 per cent of cultivated area was acquired as surplus over the ceiling limit . . . . the legislative assembly, the law court and, at the village level, a complex combination of economic and social power have worked to maintain the interests of ex-intermediaries and the dominant landholders" Dhar gives a graphic description of manoeuvres by an ex-zamindar whose estate covered three-fourths of a whole district (about 4,000 villages "rich in forests and minerals"). These manoeuvres included setting up trusts, filing of suits against the Government of Bihar, getting elected to the legislative assembly and finally, becoming a member of the select committee to which the Zamindari Abolition Bill was referred! Not surprisingly, "his membership (of select committee) enabled him to look into the bill and set up his next course of action". Interestingly, he also formed the 'Kisan Agragami Dal' which included his "relations, servants and jeth-riot . . . . (which) humbled the Congress in the district board elections".

Dhar's paper also provides glimpses of village level situations. Dhar finds that, to the extent that ryots gained from zamindari abolition, the beneficiaries were "occupancy ryots who had large chunks of land under their control". They were in the forefront in the peasant movement in the colonial Bihar. "The Bihar Land Reform Act, 1950 which came into effect in 1952 conferred proprietary rights on them and gave them unhindered authority on their surplus. In the post-independence period, they added to their holdings more land by purchases (and) by mortgages leading to sale or coercive appropriation". Dhar also mentions cases of traders and money-lenders acquiring substantial holdings and illegal occupation by the better-off sections of government lands, village common lands and lands of ex-intermediaries. Obviously, the rural poor and landless remained on the distant sidelines in the struggle for land dominated by ex-zamindars and larger ryots. The former "live under constant coercion of landowning classes . . . . (on occasions the latter) boycott him totally so that he had to leave the village and join the ranks of migrant labourers in Ludhiana".

As Pradhan Prasad has pointed out, 'semi-feudalism' tries to entrench itself by opposing agricultural modernisation and growth of capitalist farming. Consolidation of holdings is a programme which provides a favourable setting for agricultural growth. A study of this programme in Bihar carried out in 1985 finds that "out of 168.72 lakh hectares of agricultural area (consolidation work) has been completed only in 11 lakh hectares and it was frankly admitted that, out of this land, in about two-thirds of the area the cultivators have not shifted to their newly allotted chaks" (Verma, M. S., 1985). It should be mentioned here that the Consolidation of Holdings Programme was started in Bihar in 1958! According to the study, "Due to faulty implementation, dilatory completion, lack of publicity and dominance of the vested interests of some influential landowners, the programme is not popular . . . . (due to) small holdings and extreme poverty the cultivators are unable to develop their land . . . . so they are unable to take the full advantage of the programme".

A recent penetrating analysis by Pradhan Prasad looks at the question of how to get a breakthrough in the Bihar situation (Prasad, H Pradhan *et al.*, 1989). He emphasises the need for an integrated approach covering growth policies, poverty alleviation programmes and land reforms. "Strengthening and implementing of land reform policies (particularly conferring of tenancy rights on sharecroppers) will go a long way towards weakening the stranglehold of semi-feudal relations of production on rural economy in the plains of Bihar. Our results high-light the failure of these policies of date . . . . (as regards) the policies targeted for basic needs and poor (they) can provide only interim relief unless the democratic institutions get strengthened and the rural economic base is strengthened and diversified. The previous programme, for instance, would make a great deal of sense as a part of comprehensive package aimed to reduce poverty, for it reaches a group which may not benefit from programmes to increase production and employment. . . . The strengthening of the economic base would require a massive governmental effort towards increasing the area under 'well managed' productive irrigation, flood control and drainage . . . . (now) irrigation exists only in a limited area and much of it is 'protective' in nature rather than aimed at production growth". Rural electrification and generation of rural power supply is also mentioned as an essential precondition for development. Another essential precondition is "compulsory free education up to at least secondary level".

The glimpses of the land reforms situation in Bihar provided above may convey the wrong impression that the situation is hopeless and beyond the reach of the government. It should be mentioned here that some growth pockets have already emerged in Bihar agriculture. More important, as Pradhan Prasad finds, the poverty alleviation programmes do reach the poor. The main implication of the Bihar situation for land reforms is that the reforms would not succeed by themselves alone in isolation in a setting of chronic social and economic backwardness. They need to be part of a broader long-term strategy for focusing development on the poor based on a judicious combination of reform measures with growth policies and with programmes for providing immediate relief and support to the poor.

We now turn to Karnataka as an example of a situation with better land reforms performance than in the semi-feudal areas but where the combination of conditions are not favourable for a strong thrust and enduring change. A characteristic feature of the Karnataka situation appears to be that the state has initiated a number of processes of change but has not been able to sustain them at an adequate pace over a long enough period to produce a sizeable cumulative impact. For example, Karnataka's agricultural growth was quite promising up to about late 70s. There is striking evidence that some of the dryland crops and backward districts in the state benefited from growth during this period, possibly for the first time in the recent decades (Nadkarni, *et al,* 1983). However, the 80s have been the bleak years for Karnataka agriculture and there is as yet no assurance of a firm and positive trend in agricultural production and productivity over the next critical decades for the Karnataka economy. In the field of political mobilisation, Devaraj Urs had been able to achieve a notable success in bringing together the hitherto neglected groups in the state and posing a credible challenge to the political influence traditionally exercised by the two major dominant castes, viz, the lingayats and the vokkaligas. But, the process of mobilisation did not percolate down to the village levels, and, consequently, one does not find in Karnataka organisation of poor as effective and wide-spread as in Kerala or West Bengal. Meanwhile, without a strong base and organisational set-up at the grassroots level, the coalition fashioned by Urs could not build upon its initial successes and, after that passing away of Urs, the dominant castes appear to have re-emerged on the political scene in the state with little difficulty. As a third example, Karnataka is often lauded for its bold experiments in Panchayati Raj and decentralisation, particularly the radical-looking recent round launched in April 1987. And, yet, few observers to Karnataka scene appear hopeful that the district and mandal level arrangement for panchayati raj would become operative soon and fast enough to weaken in near future the prevailing constraints on rural development.

The three dimensions noted above—agricultural productivity and production, political mobilisation of poor and Panchayati Raj—are important determinants of success in land reforms. Let us describe in some detail the link between the two in Karnataka. Until the end of 60s, land reforms remained ineffective in Karnataka. While the first steps towards formulation of legislation for abolition of intermediate tenures like 'inams' were initiated in 1947, the laws reached the stage of implementation only by the mid-50s. Aziz summarises the meagre achievements of this reform as follows : "the spirit of the law appears to have got diluted because of (a) the long delay in putting the law into the statute book, and (b) the erstwhile intermediaries giving prolonged legal battles in the courts. . . . Large size of land was retained by the erstwhile intermediaries under the guise of personal cultivation. . . . sub-tenants and cropsharers did not benefit at all because they did not have the occupancy rights on the land they cultivated" (Aziz, 1990). Aziz also notes that the tenancy and ceiling legislations during this period were mild and easy to evade.

Interestingly, the 70s witnessed a quickening of the tempo of land reforms in the state. Compared to the earlier provisions for resumption of land having many loopholes which a landlord could use to his advantage, "the 1974 act which abolished tenancy clearly stated that the right of resumption was to be totally eliminated in all cases except soldiers and seamen who would be permitted to resume the tenanted land. Besides leasing of land in contravention of the law would amount to loss of land on the part of the owner" (Aziz, 1990). Similarly, there was also a bold attempt to enforce more ambitious ceiling legislation. In he first round of ceiling legislation undertaken in the early 60s, the ceiling was fixed at 27 standard acres of land for a family of five members which was declared as equivalent to 216 acres of dryland. With the provisions for extra average to additional members, a family of 10 members could reach the maximum of 432 acres of dryland! The ceiling act passed after 1972 reduced the ceiling limit to 10 to 12 standard acres though the provisions permitting retention above the ceiling (for families having more than five members, etc.) could not be completely plugged. On the whole, the second round of land reforms legislation enacted in Karnataka during the 70s was radical enough for the state to get bracketed with Kerala and West Bengal (Government of India, 1988).

It is our hypothesis that this second round of legislation became feasible because of the political mobilisation of the non-dominant backward castes and classes referred to above. It is also possible that the buoyancy of Karnataka agriculture during the 70s played a contributory role. The links between 'radical' content of land reforms on one hand and the political and agricultural environments on the other need to be explored and analysed in depth. Our purpose here is only to suggest that land reforms cannot be achieved only through good intentions but need a congenial setting. The decade of the 70s in Karnataka seems to have provided such a setting.

But, agricultural production lost its momentum in the 80s and the political wave initiated by Devaraj Urs lost its thrust. Predictably, the 'radical' content of the second round of reforms legislation in Karnataka largely evaporated during the phase of implementation. In a sense, this was consistent with the hypothesis that effective land reforms need a congenial setting since the weaknesses in implementation could have been due to the changed and noticeably less favourable setting of the 80s. What is important for us to note is that the benefits of implementation were not focused on the rural poor and landless (Table 1.2). The landless did not gain any land under tenancy legislation. Ironically, the principal beneficiaries of the tenancy legislation were the lessees having the farm size of 15 acres and above! This could have happened due to two connected reasons; the group of larger cultivators had in reality leased in more land than the smaller group and the former were in a better position to assert their rights as tenants. Equally ironically, even the ceiling legislation appears to have spread its meagre surplus available for redistribution over a heterogeneous group of beneficiaries ranging from those not owning any land to those owning up to 10 acres. Besides, the schemes specially introduced to help the allottees of surplus land were found to be "inadequate considering the number of prospective beneficiaries and the actual cost of land development" (Government of Karnataka, 1987). It may be mentioned here that findings on the impact of tenancy legislation in Gujarat were similar to the experiences in Karnataka (Patel, B.B., 1990; Sharma, Usha S., 1990).

The meagre impact of land reforms on the rural poor and landless in Karnataka occurred despite conditions more favourable for such reforms than in the 'semi-feudal' areas. Obviously, one has to look for a different set of reasons in Karnataka than those noted above in Bihar. A suggestive clue we have is that the reforms of the 70s in Karnataka got diluted at the stage of implementation rather than at the earlier stage of legislation. It would hence be reasonable to assume that it was the inability of the rural poor and landless to confront the landowning groups at the ground level which was responsible for the limited impact of the reforms on the former and the leakage of benefits to the latter. This basic handicap could have been aggravated by contributory factors such as deplorable land records, unhelpful revenue officials and a government whose zeal for radical reforms had begun to abate in the wake of changes during the 80s.

Within the spectrum of land reforms situations in India, Kerala and West Bengal form a situation which is the polar opposite of the situation in Bihar. There has been a genuine attempt in these states to formulate strong reforms and to successfully focus them on the rural poor. The reforms have been supported by adequate political commitment at the highest level and by considerable 'pressures from below' exerted by articulate and well organised groups of rural poor. However, the other—and less happy—side of the picture is that the reforms have hardly helped the rural poor to become economically viable and to overcome the socio-economic barriers keeping them on the periphery of the society, polity and economy. The following paragraphs offer glimpses of both these sides of the picture, the happy as well as the less happy. The intention is not to go into details but only to bring together insights on the curious paradox of 'successful' land reforms coexisting with the rural poor remaining peripheral to the main system and having meagre access to viable economic opportunities.

Tables 1.3 and 1.4 indicate that Kerala imposed an upper limit on holding at 6.07 hectares as compared to the very much higher upper limit between 21.85 and 24.28 hectares in its neighbouring southern states and that the number of tenants who acquired ownership of leased-in land and the extent of such area were notably higher in Kerala as compared to the performance in other southern states. More important, an analysis of distribution of operational holdings by George found that the changes between 1970-71 to 1976-77 clearly indicated the effects of land reform in reducing the inequalities in distribution (George, P. S., 1986).

TABLE 1.2 : Land Gained by Different Size Classes of Farmers under Tenancy and Ceilings Provision in Karnataka

Size Class of Farmers	*Land Gained Per House (in Acres)*	
	Under Tenancy	*Under Ceilings*
0 acres of land held	—	2.30
0-2.5	2.49	2.25
2.5-5.0	3.02	1.73
5.0-10.0	2.99	4.48
10.0-15.0	3.00	Nil
15.0 and above	9.00	Nil

Source : V. S. Satyapriya and S. Erappa, "Land Reforms in India : Some Field Evidences" in A. R. Rajapurohit (ed), *Land Reforms in India,* Ashish, New Delhi, 1984.

TABLE 1.3 : Number to Tenants Receiving ownership of Leased-in Area

State	*No. of Tenants*	*Area (in Acres)*
Andhra Pradesh (Telengana Area)	33,000	82,000
Karnataka	57,096	65,844
Kerala	2,48,429	2,40,000

Source : *Commerce,* Vol. 133, August 14, 1976

TABLE 1.4 : Ceiling on Land Holdings in Hectares (1972)

Andhra Pradesh	4.05 to 21.85
Karnataka	4.05 to 21.85
Tamil Nadu	4.86 to 24.28
Kerala	4.86 to 6.07

Source : Haque, T. and Sirohi, A. S., 1986.

Despite these apparent positive impacts of land reforms, it is doubtful that they provided any substantive benefits to the rural poor. For example, a survey of 17 villages showed that the rich peasants benefited most from abolition of landlordism (see Table 1.5). According to the findings of this survey, holdings less than 1 acre in size—which were nearly 16 per cent of total holdings—obtained less than 1 per cent of land redistributed as a result of land reforms. Holdings less than 5 acres in size (over 85 per cent of total holdings) received only a little over 36 per cent of such land with the rest—nearly two-thirds—going to holdings more than 5 acres in size accounting for only 16 per cent of total holdings.

Even when the rural poor received land as a result of reform measures, they found it difficult to hold on to it. For example, the study by George discovered that subdivision of land due to new family formation reduced the holdings of small and marginal cultivators and, further, that the rising market price of land due to the influx of Gulf remittances tempted the holders with small pieces of land to sell them. An important conclusion of this study was that the losing of land by the small cultivators led to decline in cropped area, aggravated their non-viability and impaired their investments in land improvements.

In view of this adverse environment in which the rural poor are located, the following two overall assessments of land reforms in Kerala would be particularly relevant to be taken note of by the policy-maker.

(a) "A review of some of the major institutional reforms and the economic development programmes initiated by the governments specifically aimed at the upliftment of the 'scheduled' sections in Kerala (scheduled castes and tribes) indicated that their impact has been quite modest. For instance, land reforms (have) almost by-passed them in the course of the long-drawn process of legislative deliberations and implementation . . . . the new 'landed' class (i.e., beneficiaries of land reforms from among the poor) enjoy freedom from their traditional socio-economic bondages and feudal obligations . . . . (but) the new status has not guaranteed them economic sustenance since their holdings are too small to generate even a subsistence income. The 'schdeuled' communities continue to be the most marginalised section. They still seem to suffer from disabilities despite the introduction of several programmes aimed at their upliftment" (Sivanandan, 1989).

TABLE 1.5 : Distribution of Land Released by Land Reforms by Size of Holding

Size of Holdings (Acres)	*Per cent of Households*	*Per cent of Land Gained*
0-1	15.6	0.8
1-2	34.4	13.5
2-5	34.2	21.9
5-10	11.5	31.5
10-15	2.5	7.6
15-20	0.6	10.4
20 and over	1.2	14.3
	100.0	100.0

Source : Herring, R. J., 1980.

(b) K. N. Raj, eminent scholar and proponent of strong land reforms, summed up the Kerala experience on land reforms thus "evidently, it is the improvement in the terms of trade (of its export products) and the large-scale inflow of remittances from migrants abroad that have maintained some semblance of prosperity in the state in recent years. This cannot last very long, unless the opportunity is used for strengthening the technological basis for more rapid agricultural growth in the future : the land reforms that have been implemented provide a sound institutional framework for such strengthening but cannot obviously be a substitute for it" (Raj, K. N., 1983). According to Raj, "nothing definitive can be said about the impact of agrarian reform on poverty in Kerala".

Despite some major differences in the features of their agrarian and rural economies, Kerala and West Bengal share a common land reforms scenario, viz, the reforms were successful but did little to improve the economic status and viability of the rural poor. Given the constraints of space, we give below the gist of three recent assessments of land reforms in West Bengal.

(a) "The objectives of 'Operation Barga' to formalise the relations and establish a fair exchange between the tenants and the landlords is yet to be fully realised . . . the increase in input application rate in real terms has been marginal . . . yield rate has hardly witnessed any increase over time . . . investments in agriculture and human resources have not reached any significant proportions" (Agro-Economic Research Centre, Visva-Bharathi, 1986).

(b) "The old style large-scale absentee landlordism of the early days no longer exists in West Bengal (due to) land reform legislations of 1953, 1956 and 1973 . . . (and) the call given by the United Front government of 1967 and 1969-70 to the peasants to identify ceiling-surplus land and to get these redistributed by forming committees of peasants. . . . What matters now is the role of jotedars, the former peasants who have withdrawn themselves from direct cultivation and are now the biggest landholders in the countryside and who employ sharecroppers and agricultural labourers for carrying out cultivation. . . . Whatever be their economic power, which is still considerable, the jotedars in West Bengal no longer exercise the kind of political and social authority that they did enjoy even ten years ago. Panchayats as elected bodies of rural people, kisan sabhas as a mobiliser of peasants and the left-wing political parties—all active at the grossroots level—have seriously eroded their influence over the rural population. . . . Large-scale recording of sharecroppers, near universal recording of home-steads, absence of evictions of sharecroppers and agricultural labourers, implementation of minimum wages legislation and defiance of the landlords by the poorer sections . . . are some of the major features of rural life in West Bengal today. . . . Share-croppers are no longer dependent on the landlords for loan, marketing, protection and land for dwelling houses. The situation is similar in the case of agricultural labourers . . . rural labour market is much wider and broadly-based than in the past (and) wages are standardised (and) are in conformity with the wages in similar occupations in the village and the neighbouring area . . . in the case of sharecroppers, the bulk of the production is still for subsistence and not for the market . . . given the small amount of land under their command even achievement of self-sufficiency in food production is beyond them, which is why they largely supplement their earnings with wage income. . . . Distribution of mini-kits, greater access to organised sources of credit and membership of co-operatives have created conditions for more notable growth in production in future . . . (in their absence), by the operation of market forces, control over land would soon be passed on to the richer elements . . . in the case of agricultural labourers, increases in wages and employment (due to) minimum wage legislation, public works programme along with a greater sense of security have been in evidence in all the areas. . . . A major problem, of course, is the phenomenon of growing landlessness and proletarianisation of the poorer peasants . . . this is no doubt an alarming feature which would in future have serious negative implications for both wage and employment . . . the most significant impact of the reforms has been to enhance the sense of self-respect of these poor groups and to make them believe in their own ability and organisation. This factor is perhaps more important than the immediate material achievements" (Dasgupta, Biplab, 1987).

(c) "The agrarian reform programme, currently being implemented by the government of West Bengal, does address itself to the fundamental problems facing the agrarian economy (viz., landlordism, usury, landlessness of rural poor and their need for more employment and higher wages). In the short run, these measures are both necessary and desirable. However, (their) concrete achievements so far have been inadequate. . . . If the benefits of (these) measures are to be consolidated and built upon in the longer run, conscious efforts must be made to develop co-operative or joint farming at the next stage. This is necessary not only for pre-empting a renewed process of polarisation of income and wealth, but also for facilitating the adoption of modern technology which is essential for growth" (Ghose, Ajit Kumar, 1983).

Finally, we take a brief look at the land reforms situation in Punjab, a 'Green Revolution' area marked by an extensive and growing spread of capitalist farming. The emergence of capitalist farming in Punjab is clearly reflected in the changing distribution of operation holdings in the state (see Table 1.6), changing characteristics of large holders (see Table 1.7) and changing concentration of land (see Table 1.8). It is seen from Table 1.6 that while at the all-India level the trend has been towards increase in the numbers of marginal and small holdings along with decrease in the numbers of medium and large holdings, the trend has clearly been in the opposite direction in Punjab. A similar contrast is also seen in relation to share of large holdings in total operated area (which has increased in Punjab) and average size of large holdings (which has become larger in Punjab) as compared to the all-India level decreasing trend in both these characteristics of large holdings.

Table 1.8 provides a glimpse of how the large holdings in Punjab have been able to swim against the strong current eroding the share in total operated land and average size of large holdings in most parts of rest of India. The table shows that the concentration ratio of ownership holdings remained high in both Punjab and India over the period 1971-72 to 1981-82. In fact, the concentration ratio for ownership holding was significantly higher in Punjab than in India as a whole. It is interesting to compare the behaviour of concentration ratio of ownership holdings with that of the operational holdings. The latter was noticeably less than the former in 1971-72 in both Punjab and India, the difference being particularly marked in Punjab. This suggests that, in 1971-72, the leasing of land was from large owner to a small owner (or to a landless tenant) which had the effect of moderating the concentration of land ownership. This means, in effect, that the lease market compensated to an extent for the inequitous distribiution of land ownership. Incidentally, such a lease market would have an important role to play in a situaiton like ours with meagre political will and capacity to bring about drastic redistribution of land ownership.

TABLE 1.6 : Changing Sizewise Distribution of Operational Holdings—Punjab and India

('000 Holdings)

Size Group	*Punjab*		*India*	
	1970-71	*1985-86*	*1970-71*	*1985-86*
Marginal	518	256	35682	56748
Small	260	208	13432	17881
Semi-medium	281	291	10681	13253
Medium	248	260	7932	7920
Large	69	74	2266	1929
Total	1376	1089	69993	97731

TABLE 1.7 : Comparative Characteristics of Large Holdings—Punjab and India

	Punjab		*India*	
	1970-71	*1985-86*	*1970-71*	*1985-86*
Share in total operated area of all holdings (per cent)	26.8	29.8	30.9	20.2
Average size of large holdings (hectares)	15.5	16.7	22.1	17.2

TABLE 1.8 : Changes in Concentration Ratio (Gini Coefficient)

	Punjab		*India*	
	1971-72	*1981-82*	*1971-72*	*1981-82*
Ownership holdings	0.78	0.77	0.71	0.71
Operational holdings	0.42	0.70	0.59	0.63

Note : Table 1.6, 1.7 and 1.8 are based on the data given in Singh, J.P., 1990.

TABLE 1.9 : Average Number of Parcels per Holding (Operational) in Northern States

States	*Size of Farms*					
	Marginal	*Small*	*Semi-Medium*	*Medium*	*Large*	*All Sizes*
Haryana	1.36	1.70	2.04	2.42	2.83	1.91
Himachal Pradesh	8.14	17.99	18.91	19.82	19.89	12.93
Jammu and Kashmir	14.55	5.00	6.40	4.88	4.40	11.84
Punjab	1.11	1.44	1.61	1.74	2.39	1.57
Uttar Pradesh	2.86	4.16	5.84	7.36	11.61	3.71

Source : Government of India, Department of Agriculture and Co-operation, Ministry of Agriculture, *All India Report on Input Survey,* 1976-77 (Vol. I), 1985, p 15.

However, what is more relevant for us to note is that the lease market seems to have been turning against the small owner over the period 1971-72 to 1981-82. This process appears to have had a stronger thrust in Punjab than in India as a whole. Remembering the fact that this was precisely the decade in which green revolution accelerated the emerging trend in Punjab towards capitalist farming, it is tempting to propose the hypothesis that market forces-aided by the technocratic green revolution strategy for agricultural growth which had a dramatic impact on Punjab agriculture—helped the large holdings in Punjab to exert pressures in the lease market to achieve enlargement of their operational holdings. It is likely that the large owners withdrew the land leased out by them to the small tenants and, also, influenced the small owners to lease out their land to them (to large holders) rather self-cultivate it. It is known that the green revolution technology—irrespective of its theoretical neutrality to scale—was far more accessible to large holdings than small.

If this hypothesis is correct, it would follow that the land reforms in Punjab could not reverse the powerful forces favouring capitalist farming (and, consequently, large holder) released as a combined effect of market and technocratic strategy for agricultural growth. Specifically, the reforms could not impose an effective ceiling on the size of holding, did not succeed in preventing the large owner from resuming land for self-cultivation and gave no protection to the small owner against the forces preying on their land. Needless to add, if this scenario is allowed to prevail, the access of rural poor to agricultural land as cultivators would continue to diminish. This, in turn, would mean that they would be under increasing pressure to fall back helplessly and with weak bargaining power on the market for wage-labour in agriculture. The process could progressively usher in a state in which the small owner 'voluntarily' sells his leased out land to his large tenant and the access to land becomes more and more inequitous over time. It should be a sobering thought that the conventional tenancy reforms in this situation would favour the strong and not the weak.

J.P. Singh's recent study provides revealing glimpses of weaknesses of land reforms in Punjab (Singh, J.P. 1990). Interestingly, the capitalist farmers were favoured by both the land reform programmes which were relatively well-implemented as well as those which were not so implemented. The first category includes abolition of intermediaries and consolidation of holdings. Singh's observations on these programmes are : "The programmes of abolition of intermediaries has been generally completed by the end of 1954-55 in . . . Punjab . . . abolition of intermediary interest brought the cultivators into direct relation with the state. . . . it is only in Punjah, Haryana and Uttar Pradesh (that) the programme of consolidation of holdings has been taken up seriously . . . (it) has shown good impact wherever it has been completed. It has brought in considerable increase in agricultural production, particularly through investments in private irrigation. . . . But, it has been observed that in the process of consolidation of holdings, tenants and small marginal farmers are generally losers while influential and large farmers generally receive a better deal . . . small and marginal farmers end up receiving inferior quality land . . . tenant and sharecroppers get evicted from their leased-in-land". It is obvious that having the entire holding in a few plots—rather than having it scattered all over in small bits—would be of great help to medium and large farmers to adopt capitalist farming. It can be seen from Table 1.9 that these groups of farmers had a decided advantage in this respect in Punjab and Haryana, more so in the former.

Compared to measures for abolition of intermediaries and for consolidation of holding, those for tenancy reforms and ceiling on holdings fared badly in Punjab. According to Singh, "The level of ceiling was usually very high, classification of land gave scope for manipulation, the exemption from ceiling laws were too many and the unit of application of ceiling, i.e. family was defined loosely . . . so far, the number of persons (who have) benefited from the distribution of land is only 26 thousand in Punjab" (as compared to 37.4 thousand in Haryana, 4.5 lakhs in Jammu and Kashmir and 3.0 lakhs in Uttar Pradesh). Apparently, the beneficiaries in Punjab remained small in number because each one received on an average a larger piece of land than in other northern states (nearly 4 acres in Punjab as against 3 acres in Haryana, around an acre in Uttar Pradesh and Jammu and Kashmir and between half and three-fourths of an acre in Himachal Pradesh and Delhi). As regards tenancy reforms, Singh finds that "in Haryana and Punjab, the landowners enjoy a continuing right of resumption of land upto the ceiling limit, while in other states of northern region the maximum areas that could be resumed was fixed well below the ceiling limit . . . concealed tenancies have been flourishing in the region and ejectments of tenants have been continuing." Singh also notes that fair rent has been fixed at the level not exceeding that suggested in the guidelines in Uttar Pradesh and Delhi but higher than those in Punjab (along with Haryana and Jammu and Kashmir and that "in the states of northern region the maintenance of (land) records has been fairly satisfactory in many respects but not in respect of recording the names of tenants and sharecroppers". On the whole, Punjab provides a convincing illustration of a situation where the capitalist farmers have a dominant influence on the formulation and implementation of land reforms. This did help Punjab to achieve breakthroughs in agricultural production but, clearly, at the cost of access to land of the rural poor and weak.

The following broad propositions emerge from our consideration of the typology of four situations covered in this section, viz., Bihar, Karnataka, Kerala, West Bengal and Punjab.

(i) In none of the four situations did the land reforms help the rural poor in obtaining secure and equitable access to land in the context of growing population pressure on land and keen competition for its control among the different rural strata.

(ii) The dominant rural groups—ex-zamindars in Bihar, large cultivators in Karnataka and capitalist farmers in Punjab—exert decisive influence on the processes of formulation and implementation of land reforms to divert the benefits of the reforms from the rural poor to themselves.

(iii) Even where they are politically well organised as in Kerala and West Bangal, the rural poor get only the direct and proximate benefits of land reforms but do not succeed in bringing about an enduring inprovement in their economic status.

(iv) The policy-makers usually spend considerable time and energy in improving the legislative and implementation aspects of land reforms. While these are important aspects, the crucial criterion for successful land reforms is the extent to which they form part of a cohesive development strategy capable of integrating the objective of growth with those of equity and poverty eradication. Without the support of such a strategy, no amount of tinkering with legislation and implementation would be of much help in getting effective land reforms. Keeping this point in mind, the last section of this paper makes an attempt to identify the priority tasks in the next round of land reforms which could help in the evolution of such a strategy. Before doing this, it is necessary to take a look at the recent scenario and trends in the access of rural poor to common property lands.

IV. ACCESS TO COMMON LANDS (CPRs)

Three points need to be noted about the access of rural poor to common lands. First, the household economy of the rural poor depends crucially on the access to common lands. Second, the extent and quality of the access of rural poor to common lands have deteriorated in recent years and there is as yet no credible intervention by the governments or by the voluntary organisations to reverse this trend. Considered along with the ineffectiveness of land reforms described in the preceding section, it is obvious that the rural poor are currently coping with the accumulated and growing pressures of diminishing access to both agricultural lands as well as common lands. Third, systematic and effective interventions to improve the access of rural poor to common lands need a planning-cum-policy framework for utilisation of land resources which is capable of linking the national-level objectives and policies with the micro-level schemes, progrmmes and participatory arrangements.

Published and aggregative sources of data offer little help in elaborating the first two points noted above. Fortunately, there are two recent field studies (one of which is still in progress) which have looked into the problems of access of the rural poor to common lands carefully and in detail. Let us first take up the question of dependence of the household economy of the rural poor on their access to common lands. Table 1.10 has been compiled from Jodha's study which covered 82 villages from 21 districts scattered in seven major states in the dry tropical zone of India (Jodha, N.S., 1990). The study defined the CPRs as including "community forest, pasture/waste land, pond/tank, river/rivulet, watershed drainage/river banks and river/tank beds". The study found that the rural poor (i.e., "agricultural labourers and small farm (having less than 2 hectares of dryland equivalent land) households"), derived the major part of their fuel and fodder supplies—from two-thirds to over four-fifths of the total—from the common lands. The common lands also made significant contribution to their employment (128 to 196 days in a year per household) and income (CPR income accounting for 14 per cent to 23 per cent of income from other sources). More important, there is a clear indication in Table 1.10 that inclusion of CPR income brings about substantial reduction in the inequality in income distribution. Thus, clearly, the support provided by the CPRs is important in sustaining the household economies of the rural poor.

TABLE 1.10 : Dependence of Poor on CPRs

Fuel supplies (per cent of CPR to total)	66 to 84
Animal grazing (per cent of CPR to total)	69 to 84
Per household employment (days) in a year	128 to 196
CPR income as proportion of income from other sources (per cent)	14 to 23
Gini-coefficient of income from	
(a) All sources	0.33 to 0.41
(b) All sources excluding CPRs	0.44 to 0.50

Source : Jodha, N.S. 1990.

Dependence on the CPRs is an ambiguous indicator of the development status of the rural poor. In the conext of effective policies for optimum use of CPRs and for adequate share for the rural poor in the returns from the CPRs, such dependence could be regarded as a positive indicator. However, in the absensce of effective policies which is true of the prevailing situation in India, the dependence on CPRs is more likely to reflect the precariousness of the rural poor's struggle for survival. Given the peripheral position of the poor in relation to the mainstream economy and his meagre access to remunerative income-earning opportunities, he falls back on CPRs more out of compulsion than choice. In such a situation, diminishing access to the CPR would be a disaster for the rural poor diminution in CPRs should be a point of grave concern for the policy-maker. As Table 1.11 indicates, there has been a precipitous fall in the availability of the CPRs over the period 1950-52 to 1982-84 and an equally alarming rise in the pressure of population on these resources. Taking the median values of the figures in each column of Table 1.11, it is observed that the percentage of CPRs to total village area declined from 21 in 1950-52 to 12 in 1982-84 and the number of persons per 10 ha of CPRs inceased from 46 to 117 over the same period. These trends are so sharp and clear that it is hardly necessary to comment any further on their serious implications for the future.

TABLE 1.11 : Decline of CPR Areas

State	*CPR Area as per cent of Total Village Area*		*Persons Per 10 ha of CPR Area*	
	1950-52	*1982-84*	*1951*	*1981*
Andhra Pradesh	18	11	48	134
Gujarat	19	11	82	238
Karnataka	20	12	46	117
Madhya Pradesh	41	24	14	47
Maharashtra	22	15	40	88
Rajasthan	36	16	13	50
Tamil Nadu	21	10	101	286

Source : Jodha, N.S. 1990.

As systematic utilisation of CPRs needs area-cum-village-specific planning, it would be of some interest to obtain an idea of the CPR situation in different types of villages. The data from a field-study in progress in the Institute for Social and Economic Change permit a rough comparison as between relatively developed and relatively backward villages. Tables 1.12 to 1.16 present these date. The names of the villages are withheld as our collection of data on the many delicate issues connected with CPRs are still continuing in them. The villages are arranged in the tables in the descending order to their developmet status, i.e. village A is the most developed among the four an village D the most backward. It can be seen from Table 1.12 that the availability of CPRs is negligibe in the developed villages, being less than 5 per cent of total village area. The backward villages, on the other hand, have greater extent of CPRs—between one-fourth and one-third of total village area. A clue to these differences is provided by Table 1.13 which shows that in the developed villages big farmers, dominant caste households and the better off in the village encroach on the CPRs while in the backward villages the encroachers belong to the poor and disadvantaged groups.

TABLE 1.12 : Availability of CPR Lands (Acres)

Villages	*Total Geo Area*	*CPRs Available in the Village*
A	452	15
B	923	45
C	3489	810
D	642	182

Note : The availability of CPRs shows in this table is exclusive of encroachments, CPRs taken up for regeneration or already distributed to individuals.

Source : Pasha, S.A. 1990.

TABLE 1.13 : Encroachment on CPRs

Village	*Area (Acres)*	*By*
A	10	Better off people
B	60	Dominant caste big farmers
C	350	SC, ST poor and better off
D	134	Backward Castes and poor

Source: Pasha, S. A., 1990.

TABLE 1.14 : Divergence Between Land Records and Survey Data

Village	*CPRs Available in the Village (Acres)*	
	According to Land Records	*According to Survey*
A	53	15
B	390	45
C	1011	810
D	372	182

Source : Pasha, S. A., 1990.

The process of privatisation of CPRs seems to work in the following manner. Where the rich compete for the control of CPRs, privatisation gets close to be the final limit of the available CPRs. Plausibly, the rich have the means to invest in CPRs and make them productive. On the other hand, in the backward villages the field remains open to the poor. It is likely that the rich do not exist in sizeable numbers in backward villages or do not have a command over resources comparable to that of their counterparts in the developed villages. One implication of such process of privatisation is that the development potential of CPRs is likely to be much larger in the backward villages than the developed villages. At the same time, it is also reasonable to assume that the exploitation of this potential needs resources—technology, expertise, investments—which are beyond the capacity of the local communities to provide. Another relevant insight arising from our findings is that, contrary to the widely prevailing view, it is the rich rather than the poor who are mainly responsible for the privatisation of CPRs. If this process is allowed to operate unhindered, the poor will continue to lose their access to CPRs or get pushed into those parts of CPRs which have too low status to attract the notice of the rich. It is the same process which in the first instance made the poor landless in agriculture and is now eroding their access to CPRs.

Reversing this process needs both a comprehensive policy framework at national level and a series of area-specific schemes and programmes at the grassroots level. The first priority needs to be accorded to measures which could clear the ground for systematic and sustained policy interventions. A preliminary measure of utmost importance is to improve the land records. We found in the four villgaes referred to in Tables 1.12 and 1.13 that there was wide divergence between the availability of CPRs according to land records and the availability as verified in the villages (Table 1.14). Significantly, the divergence is seen to be the widest in the developed villages marked by the encroachment of the CPRs by the rich. Obviously, the divergence arises not so much from errors as from a design to conceal. It may be mentioned here that the Government of Karnataka set-up a group consisting of top officials some four years back (in early 1986) to prepare a village-wise inventory of wastelands by types. In mid-1990, the work was still in progress! Needless to add that without a reliable data base it would be idle to plan policies and programmes for CPRs. Second, a careful assessment needs to be made of the investments needed to get adequate returns from CPRs. As can be seen from Table 1.15, in the four villages covered in our Survey, the CPRs did not include village forests and only a small part of them was Gomal land. Over 80 per cent of CPRs are C and D lands in an extremely degraded conditions.

Third, it has been repeatedly pointed out that the CPRs in India lack a management system for their protection, conservation and rational use. As one observer puts it, "The central and state governments are holding, together with village panchayats, most of India's common resources. However, many resources are hardly managed and (are) in practice open-access resources", (Krooneman, 1990). Jodha found that in all of his 82 sample villages, excepting two, there were formal or informal regulations for CPR use prior to 1950s but by the early 1980s when he visited the villages only in eight of the 82 villages the regulations were still operative! (Jodha, 1990). In a word, the pre-condition required for policy intervention in the area of CPRs is the building up of management system at the level at which (district, block or mandal) the Planning Commission proposes to organise decentralised planning (Planning Commission, 1990(b)). The management system should compile and periodically update a reliable data base and also arrange for a competent technical evaluation of the potential use and investment requirements of CPRs.

In the absence of a sound management system, the schemes now being taken up on CPRs tend to be *ad hoc* and directionless in nature. Table 1.16 shows that distribution of CPRs to individuals as house-sites or for cultivation accounts for nearly two-thirds of CPR area covered by programmes.

Ironically, not many of the allottees carry out cultivation in practice. "The beneficiaries usually (in 60 to 90 per cent cases) have to sell the land as they do not have enough resources to cultivate the marginal land" (Jodha, 1985, 1986). Very likely, the allotted lands also are too poor to make cultivation profitable. Compared to allotment to individuals for cultivation, the schemes for regeneration of CPRs cover less areas. Table 1.16 also shows that the voluntary agencies play a limited role in these schemes. Schemes planned and operated by the collective efforts of the local people themselves are even rarer and exceptional and do not figure in our survey data at all.

It is instructive to see the chasm between the present use (or, more appropriately, neglect) of CPRs and their impressive development potential. According to a recent assessment, rural India presents the picture of a "cruel paradox of mass poverty coexisting with vast resource potentials. . . . These potentials were seen to lie in changing resources access and use and included canal irrigation, lift irrigation, land reforms, wastelands development, forests and agro-forestry . . . (these resources) presented large-scale potentials which had not yet been exploited. We found evidence that key aspects of (these resources) had been neglected and that the scope for benefits to the poorer were greater than commonly supposed" (Chambers, R., 1989). A point which is often overlooked is that the neglect of CPRs is part of the larger problem of lack of a national-level policy framework for conservation, upgradation and optimum utilisation of land and water resources in India. Eroded soils, silted up tanks and reservoirs and land lost to cultivation due to salinity and waterlogging are the obvious consequences of the absence of a comprehensive policy framework for the land and water resources. Declining standards of cultivation in dryland agriculture, its inability to retain high-valued crops (like cotton) and the increasing precariousness of the economy of small and marginal farmers are other serious consequences attributable to the same reason (Reddy, A.M. 1990 and Reddy, V.R., 1990). A particular mention needs to be made here of the contribution which CPRs make to the economy of small and marginal farmers. As Table 1.17 shows CPRs are important for these farms in mobilising inputs, putting the available land under crops, developing crop by-products and surviving through crisis like droughts.

TABLE 1.15 : Distribution of CPR Lands by Categories

(Acre)

Village	*Forests*	*Gomal*	*C and D*	*Tank Bed*	*Other*	*Total*
A	—	15	—	—	—	15
B	—	10	30	—	5	45
C	—	—	750	60	—	810
D	—	80	100	2	—	182

Source: Pasha, S.A., 1990.

TABLE 1.16 : CPRs Covered by Programmes

Village	*Projects of Regeneration*	*Distributed to Individuals*
A	15 (Gomal by a voluntary agency)	5 (to SSs for housing)
B	50 (Gomal by forest department)	225 (Darkhast)*
C	42 (C and D by forest department)	50 (social security plantation for SCs)
D	50 (Gomal by forest department)	—

Note: * Land allotted to SCs and poor for crop cultivation based on the applications made by them. This practice has now been stopped. Dark hast land also includes regularisation of encroached lands.

Source : Pasha, S. A., 1990.

TABLE 1.17 : Contribution of CPR to Farming Systems

CPR components in total own inputs (pre-sowing to preharvest) per cent	31 to 42
Potential decline in the absence of CPR	
Draft power	68 to 76
Manure/dung	35 to 43
Area under crops	48 to 55
Crop byproducts for sale	84 to 96
CPR contribution to total sustenance	42 to 57
Income during drought years	

Source : Jodha, N. S., 1990

While the deterioration in CPRs and diminishing access to them affect all the rural poor, tribals form a group whose very existence as a distinctive socio-cultural entity comes under intolerable strain. According to the 1981 census, scheduled tribals had a population of 54 million or nearly 8 per cent of the total population of the country. Out of these, nearly 48 million are forest dwellers living within or in the vicinity of forests (Swamy, M. C. K., 1990). In addition, several rural settlements exist in the vicinity of forests, arid or semi-arid regions, hilly terrains and foothills of the Himalayas whose dependence on CPRs for the economic and livelihood patterns is substantial. Swamy points out that "nearly 80 per cent of lands of tribal households are communal land as well as other parcels of land that are owned or held by the government." Policies for forest and wildlife conservation need to be formulated keeping in mind the interests of tribal population. The communal systems of land-ownership among the tribals also need protection against marketisation of land and its alienation to non-tribals.

V. IDENTIFICATION AND CLASSIFICATION

Currently, the rural poor are confronted with a grim situation. The livelihood opportunities available to them are hardly keeping pace with the increase in their access to land—agricultural and common—has already been indicated in the preceding sections. Given the modest agricultural growth, the growth of rural non-agricultural activities also remains low. Consequently, the only avenue left to the rural poor is to seek wage employment in agriculture. Not surprisingly, the working paper prepared in the Planning Commission on the employment situation on the eve of the Eighth Five Year Plan shows a number of disturbing signs—more open unemployment, increasing casulisation and growing dependence on unreliable and low productivity employment in the unorganised activities (Planning Commission, 1990(a)). In these circumstances, the credibility of land reforms as a development strategy needs to be assessed chiefly in terms of the contribution they can make to the viability and welfare of the rural labourers.

It would be obviously unrealistic to assume that all the rural poor can be accommodated in agriculture as owner-cultivators. The strategy capable of making the poor viable has necessarily to be multi-dimensional. This means that land reforms have to be a component in a package of measures specifically tailored to the circumstances of rural labourers. In the following few paragraphs, we briefly describe three packages—more to illustrate the idea of a package than as definitive prescription for the problems of rural labourers. It is useful for this purpose to distinguish three groups of wage-paid labourers. The first group consists of those who have always been labourers having never had cultivation as an occupation in their living memories. Asking them to become cultivators may not be the right strategy to meet their problems. We would put in the second group those who, before joining the ranks of labourers, were small/marginal cultivators but either decided to move out of farming (owing to meagre returns in farming) or were forced out by circumstances beyond their control. They are likely to have good potential to become enterprising cultivators provided they are helped with adequate incentives and supporting measures. The third group would include those who are cultivators but turn to wage labour for supplementary income. It would seem reasonable to assume that these three groups would need different packages of measures to help them towards achieving viability.

The package suggested for the first group is the following. Wherever possible, we have indicated the present status of the component and the need to strengthen it.

Employment Programmes

This component is needed to support incomes and to remove the excess supply in the labour market. According to the sub-committee of the parliamentary consultative committee which prepared a report on the problems of unorganised workers in agriculture sector, the employment programmes would have to be increased more than eight-fold over their present level if the modest target of providing 100 days of employment in a year to at least one member from each household below poverty-line is to be achieved.

Enforcement of Minimum Wages and Social Security

This is an essential complementary measure to employment programmes. The sub-committee mentioned above observes "It can be safely said that nowhere in the country minimum wage for the agricultural workers is enforced round the year". The necessity of revising the minimum wages at regular intervals has also been neglected so far. As regards social security, only a couple of states seem to have made a beginning.

Allotment of House-Sites and Janata Houses

Provision of house-sites to agricultural labourers should be cosidered an important item in land reforms. House-site and house not only meet a basic need but also enable households to take up supplementary land-based activities like vegetable growing, poultry, dairying, etc.

Social Forestry/Tree Pattas

Another component of land reforms relevant to labourers is imporved access to common lands to add to their incomes. According to a recent estimate, the tree pattas granted up to 31-12-1988 were 1.1 lakh in number covering 2.52 lakh acres (National Commission on Rural Labour, 1990). Needless to add that this is no more than a very modest beginning of a programme which needs far more attention and priority in future.

The package suggested above is based on the strategy of integrating the measures for employment generation and impovement in working conditions along with giving at least a modest land-base for labourers. This would not only add to the income-earning opportunities available to them but also reduce their precarious dependence on employers and wage-labour.

We would strongly urge that the package appropriate for the second group (viz, potentially enterprising cultivators who are now landless) would be the one based on integrating the first three components of the package described above with the distribution of surplus land. The group should be helped over time to become viable cultivators and to reduce dependence on wage-labour and employment programmes which could be reserved progressively for the first group (i.e. those who would continue to depend on the wage-labour to a significant extent even in the long run). This is also true of the access to common lands. On the other hand, not only should the poor in the second group be given priority in the distribution of surplus land but, care must be taken to give them a holding above a certain minimum size to ensure viability. They should also be assisted with credit, extension, inputs, etc. on an adequate scale. The past programmes for redistribution of land to the landless have been ineffective due to allotment of areas below the threshold of viability and inadequate assistance of allottees. These errors ought to be avoided in future.

Regarding the third group, the strategy should be to integrate land reforms with the programmes of effective use of land. Looking back on the post-independence land reforms, a striking point which emerges is that the concern has been much more with equity (mostly in the form of unimplementable rhetoric) than with equity-and-productivity. The result has been that the land reforms have achieved neither of these objectives. The main links between equity and productivity which need urgent attention in the next phase of land reforms are the following.

Consolidation of Holdings

This programme has been undertaken only by a few states in the north. The background note on land reforms prepared by the Planning Commission in 1989 mentions that 140 million acres of land have been consolidated by that year. If this figure is true, it would mean that about 40 per cent of total cultivated land has been consolidated. However, it is doubtful that the figure of 140 million acres reflects the ground level reality for the simple reason that not much is known about the strictness with which the measures for prevention of fragmentation are formulated and enforced. In the absence of these measures, the benefits of consolidation would not be retained over time. It is necessary to emphasise that among the different types of land reforms consolidation of holdings is possibly the most important measure capable of bringing about substantial improvement in the productivity of land.

Land Records

An expert report for the Planning Commission has observed "In all parts of the country, records relating to land are in a very bad shape. The record as it exists today hardly reflects the present-day reality regarding ownership of land . . . disharmony between the record and the reality not only destroys the utility of the record but also misleads any person who has to deal with land. . . . The preparation and maintenance of record-of-rights in land got low priority in all the seven Five-Year Plans" (Wadhwa, D.C. 1989). It is obvious that we have hardly any system—leave alone an effective system—to monitor land ownership, transfers and lease operations. More important, it is inconceivable that in the absence of reliable land records it is possible to have a satisfactory inventory of land resources, effective programmes for their conservation and upgradation and planning for their fuller use. It is instructive to contrast this attitude with the empty rhetoric about imposition of 'radical' ceilings on holding. If we had an adequate monitoring and planning system for land resources, the poor—particularly the poorest strata like labourers—could have gained much more from land reforms which, in addition, would also have had a substantive impact on the value and productivity of land resources. Important consequences of absence of good land records are indicated below.

Equity-cum-Productivity

(i) Ineffective imposition of ceilings on holding owing to concealment of surplus land;
(ii) Emergence of 'informal' tenancy on a large scale having the effect of exploiting the tenants and hindering modernisation of cultivation;
(iii) Continuation of fragmentation of holding undoing the benefits of consolidation programmes;
(iv) Negligent use of land—particularly dry land—in large holdings held by owners disinterested in farming as an enterprise;
(v) Privatisation of the common lands by the rich which means their loss to the pool available to the poor and their use by the norm of profits to the rich rather than utility to the village community as a whole and its major segment, viz. the rural poor; and
(vi) Lack of attention and investments to make full use of common lands in a socially optimal way.

The three packages of measures described above have the limited purpose of illustrating an important point. The point is that even good land reforms—unless they are part of carefully tailored packages—run the risk of inefffectiveness in the Indian environment. Our land reforms suffer from the further handicap that they have been far from good. They were motivated by vague notions of equity which were more talked about than seriously implemented. More important, they were grossly ignorant or negligent about the important dimensions of use of land resources affecting productivity of agriculture and steady and sustainable agricultural growth over the long run. The next phase of land reforms would have to be planned from the structural-reformist perspective which has been ably analysed and elaborated in the recent researches of Joshi (Joshi, P. C., 1981, 1986, 1987). It is tempting to quote Joshi at some length to bring out the essential ingredients of the strucutral reformist perspective of land reforms :

> "Since independence the old semi-feudal agrarian set-up characterised by the landlord-tenant nexus has been eroded. What has replaced it, however, is a dualistic agrarian pattern putting the large owner/farmer in a dominant position and the small and marginal peasants and labourers in dopendent position. This dualistic structure now acts as a formidable constraint on the process of agricultural modernisation. It does not allow the vast mass of small and tiny producers to get involved iin the modernisation process and benefit from it in an enduring and self-sustaining manner. In many areas the modernisation process within the dualistic structure is even a destabilising force for the small peasant economy . . . (we need) a new agrarian pattern which provides incentive to the small peasants and a new organisational pattern which is sensitive to peasants' needs and problems. . . . To generate relevant modern technology and to ensure its absorption in the agricultural sector (having large numbers of) small peasant producers is a highly complex difficult undertaking. . . . Transforming peasant agriculture involves social engineering. . . . Recent experience has shown that the tendency to view the land policy purely as an operational or administrative question devoid of any philosophical and moral significance has led to disastrous consequences. . . . It is in this background that we have to consider an overall community appraoch to agricultural development, encocompassing vast masses of small cultivators . . . the new phase calls for a two-fold strategy—firstly, of containing and disciplinig the large producer in the direction of welfare-oriented growth and, secondly, of firmly protecting and assisting the small producers and labourers of participatory development. Without a restructured agricultural development strategy firmly based on the vast working peasantry, assisted by a community approach and community institutions, the target-group-based anti-poverty programmes alone cannot realise the goal of growth with social justice".

Improvement in Land Records

For identifying the priority tasks in land reforms, we have kept in view the past experiences brought out by the researches on land reforms as well as the perspective for future noted above. These priority tasks needs to be taken up with a sense of urgency and determination to implement them within a reasonable time frame.

The work of preparation of land records for the states not having them now should be completed expeditiously. The background note on land reforms prepared by the Planning Commission mentions that the areas in the north-east do not have land records.

The interim report on 'Registration of Title to Land' submitted to the Planning Commission by D. C. Wadhwa has recommended the adoption of 'Torrens System' on an experimental basis in some selected districts where the revisional surveys have been completed recently. This system permits "the conversion of the present presumptive titles to land into conclusive titles". Wadhwa has also recommended that there should be wholesale revision of all pre-1950 surveys and that along with updating present record-of-rights the system of registration of title to land may be introduced. These recommendations should be implemented with the priority they deserve.

When updating the records, representatives of rural labour and poor should be consulted to obtain reliable information on (i) ownership of land above ceiling, (ii) encroachment of common lands. It is of utmost importance that the updated records reflect the true position in the field in this respect.

It is reported by the technical committee set-up by the department of rural development, which looked into the pilot projects for computerisation of land records, that the total cost of computerisation for 450 districts (including aerial photography units) would be Rs. 150 crore i.e., Rs. 30 crore per annum in a five-year time frame. According to the committee "technology is now available to facilitate easy access to information, expedite survey operations, facilitate quick updation of land records and enable their storage and retrieval". After careful appraisal of costs and feasibility, a time-bound programme may be framed for computerisation of land records.

There should be an effective monitoring system, at the panchayat level, of lease, transfer and alienation of land by the target group households, by large owners and by non-cultivators (including non-resident owners of land in the village).

Ceiling on Land Ownership

Concerted efforts need to be made to locate (i) land held evading ceiling provisions, (ii) encroachment of common lands, and (iii) land left fallow/unused for long duration by the large owners/non-cultivators.

Appropriate legislation must be enacted to take over these lands without compensation or at nominal compensation. This is also true of good cultivable lands diverted by a large owner to other uses to avoid the problems in getting them cultivated by tenants/labourers.

Farm leaders complain that frequent downward revisions in ceiling creates uncertainty and disincentives for enterprising farmers. There is an element of truth in this complaint as any further reduction in ceiling by itself, will not yield much surplus but could hinder investments in agriculture. It is our feeling that it would be desirable to eliminate the uncertainty about the ceiling limit. It would be better to pay more attention to loopholes (higher ceiling limits for larger families, separate provision for adult sons, etc.) which need to be closed and to acquisition of categories of land mentioned above.

The imposition of ceiling on ownership of agricultural land needs to be part of a broader policy of ceiling on assets and incomes in all sections of society.

It is recommended that the distribution of surplus land should be guided by a composite criterion based on the need of the recipient as well as the objective of ensuring viable use of allotted land.

Regulation of Tenancy

The experience so far shows that complete banning of leasing of land is neither feasible nor necessarily desirable. It should be noted that even West Bengal has opted for giving protection to share-croppers rather than eliminating the system of tenancy altogether.

A more practicable course would be to insist that all lease transactions must get properly recorded and that their terms and conditions are fair to the weaker party to the transaction—lessees (in most areas) or lessors (where reverse tenancy prevails).

It would be desirable to have appropriate tax provisions on land to prevent the emergence of large and dominant group of rentiers who own land chiefly to lease it out for rent or for speculative gains. If land records are reliable, such indirect measures may be more effective in controlling dysfunctional ownership of land than direct measures like banning of land sales or leases. However, when imposing such measures, care must be taken to protect the interest of small owners who find it convenient to take up occupations other than cultivation. Even for such owners, it may be possible to devise financial assets conferring all the advantages which a small owner derives from owning land without cultivating it, viz. security access to credit etc. As a long-term objective, our policies should be to encourage and enable the actual cultivators to become owners of the land they till. But it would not be advisable to use only direct controls to achieve this objective. The instruments used should also include indirect incentives and disincentives designed to serve the same purpose as the direct controls but more effectively and at a lesser cost.

Consolidation of Holdings

The researchers who have evaluated consolidation of holdings programmes find that the main reason for the extremely slow progress of the programme is the virtually total neglect of the programme at the highest policy-making levels. There should be adequate recognition at these levels that consolidation of holdings (and prevention of fragmentation) is an essential component in land reforms in the sense that without it the other components like imposition of ceilings and regulation of tenancy would not be able to achieve the principal objective of land reforms, viz. creation of a modernised and viable peasantry.

What is important for speedy implementation of consolidation programmes is the preliminary groundwork to obtain the willing co-operation of all the strata in the community. The advent of decentralised and participatory planning set-up at the level of villages could be a powerful factor to promote such programme which have suffered in the past because of excessive dependence on bureaucratic formulation and implementation of development schemes.

It is our feeling that there would be a definite advantage in undertaking consolidation of holding schemes along with revision of past surveys, updating of land records, bringing them closer to the field realities and modernisation of system of records. It would be easier then to obtain the co-operation of target groups which are often afraid that they will get a raw deal in consolidation schemes. This is particularly true when they are associated with updating of land records, preparation of consolidation scheme and, more generally, with formulation and implementation of development programmes. Once the target groups are convinced that they can benefit from a programme, it would be easy to obtain their full and willing co-operation.

Optimum Use of Land Resources

The effort now in progress to evolve practicable methodologies for planning the optimum use of land resources need to be vigorously pursued. Agro-climatic regional planning exercise of the Planning Commission and the Watershed approach for conservation and fuller use of land resources which is being experimented with in some of the states (like Karnataka) should be developed and extended to the point where they become replicable and implementable on a wide scale. There are also other models proposed by experts which deserve to be carefully considered (Ravindranath, 1990).

Purely technocratic planning of land resources would not be of much help. For example, a recent evaluation of a project proposal from the Karnataka Forest Development Corporation for development of 2,000 hectares of C and D class lands carried out by the Karnataka State Council for Science and Technology found that drastic revisions in the proposal were necessary to protect the interests of the target groups in the area of project (Karnataka State Council for Science and Technology, 1989). The project proposal had not paid much attention to the question of impact of the project on the local target groups.

The decentralised planning at the district and panchayat levels should accord the highest priority to preparation of time bound proposals for developing common lands for the benefit of rural labourers and poor. If necessay, private enterprise may be invited to develop common lands through large and lumpy investments and modern technologies provided they agree to hand over a specified portion of developed land to the distributable pool of land available for allocation to the target group households.

Planning and Monitoring of Land Reforms

At the level of the lowest decentralised planning unit (block, mandal or gram panchayat as the case may be), a listing needs to be made of the households of rural labourers and other target groups. A household-by-household scrutiny should be made to formulate packages of measures (land reforms plus other measures as explained above to help these households. Annual targets should be laid down for covering these households. Currently, we have only scheme-wise monitoring of implementation and impact. In spite of this, it is known that the extent of corruption, leakages and ineffective implementation is very large. Once the strategy of package of measures is adopted, the monitoring will have to pay attention to the effectiveness of the package as a whole over and above observing the progress in the implementation of individual components' in the programme.

A system must be devised by the decentralised planning authority to carry out annual surveys on a sample basis to keep a watch on and to adjust the package of measures to the factors giving rise to poverty, landlessness and deprivation. It needs to be mentioned here that even in a state like West Bengal there are groups of poor who continue to be neglected (Sah, K., 1990).

It is of utmost importance to associate the leaders and representatives of the target groups while operationalising the steps suggested above. There should be periodic assessment at the state level of the progress achieved by the different decentralised planning authorities in implementation of land reforms. Suitable incentives and penalties may be devised to put pressure on these authorities and to foster among them a genuine sense of commitment to achieving the objectives of land reforms within a definite timeframe. Anticipatory action needs to be taken to protect the decentralised planning authorities from domination by the rural elite.

There should be annual meeting of National Development Council and Interstate Council with a session exclusively set apart for review of implementation and impact of schemes for land reforms and for development of common lands.

Links with other Policies

At the level of overall development strategy, land reforms need support from the broad spectrum of agricultural policies focused on improvements in agricultural investments, research, extension, credit, services and institutions. These policies should have adequate thrust to help the large number of peasants having only a modest holding.

The ultimate success of land reforms depends on the emergence of organisation of rural labourers and poor who at the moment are neither economically viable nor politically mobilised. These organisations need to operate from the grassroots level of vllage and its target group households to the highest level where the development objectives, priorities, strategies and policies get determined. Absence of a sufficiently strong hobby at the national level is a major problem now confronting the rural labourers and poor. There should be a careful study of the organisations—trade unions, co-operatives, group farming, etc—which have succeeded in the past in mobilising the poor for political and developmental action. In the light of this study, the government should devise a set of promotional measures to help the emergence of such organisations. However, there should be no government interference or domination in their working. The promotional measures should have specific provisions to counter the opposition of rural vested interests to the organisations of rural poor. It should be remembered that a good system of land reforms is one in which the tasks of building up political pressures for reforms and their formulation, implementation and monitoring are progressively taken over by the organisations of poor themselves.

Note: Reproduced from *Economic and Political Weekly,* June 27, 1992 with kind permission of the publisher.

REFERENCES

Aziz, Abdul (1990), 'Land Reforms in Southern States : Policy Legislation and Implementation' (paper prepared for the Study Group on Land Reforms and Common Property Resources, Institute for Social and Economic Change, Bangalore).

Agro-Economic Research Centre (Visva Bharati) (1986). *An Evaluation of the Programme of Barga Operation in West Bengal,* Santiniketan.

Alagh, Y. K. (1987), 'Employment and Structural Change in Indian Economy' in *Human Resource Planning : The Asian Experience,* Rashid Amjad (ed), ARTEP, International Labour Organisation, New Delhi.

Bandyopadhyaya, N. (1989), *Some Comments on West Bengal Experiences in Tenancy Reforms in Land Reforms — A Retrospect and Prospect,* Planning Commission, Government of India.

Bhattacharya, M. G. (1989), 'Status of Tenancy and Land Holdings in India' (draft report, National Sample Survey Organisation, New Delhi).

Chambers, R. *et al* (1989), *To the Hands of the Poor : Water and Trees,* Oxford and IBH, New Delhi.

Consultative Committee of the Ministry of Agriculture (1986), *Indux of the Material on Land Reforms.*

Damle, C. B. (1989), 'Land Reforms Legislation in Karnataka : Myth of Success', *Economic and Political Weekly,* August 19.

Dasgupta, Biplab (1987), 'Monitoring and Evaluation of the Agrarian Reform Programme of West Bengal' (mimeo), Calcutta.

Dhar, Hiranmay (1990), 'Institutional Constraints to Land Reform in Bihar' (paper prepared for the Study Group on Land Reforms and Common Property Resources), Giri Institute of Development Studies, Lucknow.

Dore, R. P. (1959), *Land Reform in Japan*, Oxford University Press, London.

Economic Advisory Council (Government of India (1990), *Interim Report of Employment Oriented Strategy of Development.*

George, P. S. (1986), 'Emerging Trends in Size Distribution of Operational Holdings in Kerala', *Economic and Political Weekly*, February 1, 1986.

Ghose, Ajit Kumar (1983), 'Agrarian Reform in West Bengal; Objectives, Achievements and Limitations' in *Agrarian Reform in Contemporary Developing Countries*, Ajit Kumar Ghose (ed), Croom Helm, London.

Gill, S. S. (1990), 'Organisation of Rural Labour in Northern India with special reference to Punjab' (unpublished paper), Department of Economics, Punjabi University, Patiala.

Government of India, *Agricultural Censuses* 1970-71, 1976-77, 1980-81 and 1985-86.

Government of India (1982). *Rural Tenancy in India*, Ministry of Rural Development.

Government of India (1988),'Land Reforms' in Occasional Papers (First Series), Department of Rural Development, Ministry of Agriculture.

Government of Karnataka (1987), *Report on the Working of Land Reforms Act Follow Up Measures*, Bangalore.

Haque, T., *et al.*, (1986). *Agrarian Reforms and Institutional Change in India*, Concept.

Herring, R. J. (1980), 'Abolition of Landlordism in Kerala—A Redistribution of Privilege', *Economic and Political Weekly* (Review of Agriculture), Vol, 15, No 26.

Jodha, N. S. (1985), Population Growth and the Decline of Common Property Resources in Rajasthan, India', *Population and Development Review* 11, (2).

——(1986), Common Property Resources and Rural Poor in Dry Regions of India', *Economic and Politiical Weekly*, Vol XXI, No 27.

——(1990), 'Rural Common Property Resources : Contributions and Crisis' (Foundation Day Lecture delivered on May 16), Society for Promotion of Wastelands Development, New Delhi.

Joshi, P. C. (1975), *Land Reforms in India : Trends and Perspectives*, Allied Publishers, Bombay.

——(1981), 'Problems of Land Reforms in Second Stage', *Man and Development.*

——(1986), Institutional Dimensions in Agricultural Planning : Reflections on the Mahalanobis Approach, in D. K. Bose (ed), *Review of the Indian Planning Process*, Indian Statistical Institute, Calcutta.

——(1987) *Institutional Aspects of Agricultural Development*, Allied, New Delhi.

Karnataka State Council for Science and Technology (1989), 'Impact Assessment for Karnataka Tea Project' (Working Group Report — Chairman : M. V. Nadkarni), Indian Institute of Science, Bangalore.

2

LEASE MARKET, POVERTY ALLEVIATION AND POLICY OPTIONS

G. PARTHASARTHY

There has been considerable shrinkage in the lease market since independence. The steepest decline was in the 1950s following the tenancy legislation. Area under tenancy declined from 20.34 per cent as per the NSS Eighth Round (1953-54) to 10.70 per cent in the NSS 17th Round (1961-62). The 1960s did not record any significant change in the area under tenancy. The Green Revolution, coupled with the more radical postures in land reform, in a early 1970s resulted in large scale resumptions for self-cultivation. Area under tenancy declined from 10.57 per cent in the 26th Round (1971-72) to 7.18 per cent in the 37th Round (1982).

The decline in tenancy was due to the acquisition of ownership by tenants only in a few areas, such as Kerala (Herring 1983 : Oomen 1990) consequent upon effective and successful implementation. In a few other states, such as in Maharashtra and Gujarat under the Bombay Tenancy and Agricultural Land Act, 1948 and 1956[1] (Appu, 1975); in Tanjavur district in Tamil Nadu (Rajagopalan and Anuradha, 1984); and in Telangana, following the Telangana peasant struggle in the 1950s (Khusro, 1958) some tenants acquired ownership. But in these states, except in Kerala, the tenants who gained belonged mainly to the upper strata of tenant groups, and poorer tenants, gained very little from tenancy legislation[2]. Badly implemented tenancy legislation, evictions and resumptions for 'self-cultivation' reduced the access of the marginal farmer to land (Swamy, 1988). Several writers point our growing trends of reverse leasing (Sheila Bhalla, 1983, Iqbal Singh, 1989; S. S. Gill,

1989) and emergence and growth of 'capitalist tenancy' as distinguished from 'subsistence tenancy', further restricting the access of marginal owners and the landless to the lease market.[3] Some studies (Sheila Bhala, 1983) also point out increasing leasing out by marginal owners as the holdings became non-viable in the context of new technology.

Against such a backdrop of observed trends in the magnitude and changing character of tenancy we examine in this paper:

- the extent of which access to land through reliance on the lease market continues to be relevant to the landless and marginal owners in particular;
- the significance of an effective tenancy reform in poverty alleviation programmes; and
- the policy options in making tenancy reforms effective against the context of Draft Agrarian Policy, 1990.

The empirical facts on the lease market are drawn mainly from NSS 37th Round Data. Agricultural Census Data, though available, are not considered useful since these are compiled from revenue records in which rights of tenants are not recorded. It is recognised that even NSS data, which are collected from individual households, could have a downward bias, especially in respect of leased-out areas of the medium and large ownership categories.[4] Yet it is used because it is the best available level.

All-India Dimensions of Tenancy

At the all-India level, there were 17.4 million rural households reporting leasing in (Table 2.1). These formed 18.55 per cent of total rural households, which were estimated at 93.9 million as in 1982. the distribution of households reporting leasing-in by households category showed large concentration in the landless, marginal and small owner category. These groups accounted for 91.55 per cent of the total tenants in rural India.

At the all-India level, those who lease-out land were 5.2 million rural households: most of them small owners. Some of them do not have bullock power. Some have fragments which it is convenient to lease-out. A few do not have manpower. A few others have other occupations which could be profitably pursued by leasing out. Yet what is important to note is the numbers of small leasing-out are far less than the numbers among the small who seek improved access to land

TABLE 2.1 : Distribution of Estimated Rural Households by Ownership, Leased-in and Leased-out: All-India : 1982

(In Hundreds)

Ownership Category (h)	*Total owner Households*	*Leasing-in Households*	*Leasing-out Households*	*Percentagein Total*			
				By Ownership	*Leasing-in*	*Leasing-out*	*Net Leased-In*[b]
1. Landless	10,637,1	6,113,4	123	11.33[a]	30.52	0.24	40.92
2. Marginal (0.002-1.00)	51,915,7	10,266,6 (4.01)	2,558,9	55.31	51.25	49.94	51.69
3. Small (1.01-2.02)	13,794,6	1,958,8 (1.77)	1,104,9	14.70	9.78	21.56	5.73
4. Semi-medium (2.03-4.04)	10,118,5	1,136,0 (1.19)	957,1	10.78	5.76	18.68	1.20
5. Medium (4.05-10.12)	6.050,6	497,2 (1.33)	373,0	6.45	2.48	7.23	0.83
6. Large (10.13 and above)	1,338,3	61,8 (0.52)	117,8	1.43	0.31	2.29	–0.38
	93,854,8	17,391,3	5,190,2	100.00	100.00	100.00	100.00

Notes : Figures within brackets are ratios of leasing-in to leasing-out.

The totals do not add upto grand total of the individual items because of the possibility of households reporting multiple terms.

The sources of data for this and the succeeding tables, unless otherwise mentioned, are, *Sarvekshana,* Vol. XI, No. 2, Issue No. 33, October, 1987.

a Landless includes those who own less than 0.002 h of land.

b It is possible that some households may be leasing-in and simultaneously leasing-out.

by leasing-in. For every resident owner who leases-out there are more than three resident owners who lease-in. There is also an inverse relationship between size and the ratio of households leasing-in to leasing-out. Among the marginal owners, the ratio is 4.01 while the corresponding ratio for the large owners the ratio is 0.52. It is worth recognising that among the net leasing-in households more than 90 per cent are comprised of landless and marginal owners households (Table 2.1).

In brief among those who lease-in, the bulk are landless and marginal owners. This picture does not change even if we consider those who lease out among the small. The landless and the marginal owner seek improved access to land.

Leased-In and Leased-Out Area

Out of a total estimated owned area of 119.8 million hectares, 8.9 million hectares are reported leased-in (Table 2.2). In terms of area, one-half of the total area leased in is accounted for by the landless and the marginal and another 20 per cent by the small owner group. Thus the bulk of the leased-in area, as in the case of households, is accounted for by the marginal and small owners.

TABLE 2.2 : Estimated Area Owned, Leased-in and Leased-out by Size Group All India: 1982

	Size Group	*Owned Area*	*Leased-in Area (in 00 h)*	*Leased-out Area*	*Percentage of Total*		
					Owned	*Leased-in*	*Leased-out*
1	Marginal including landless	14,862,6	4,115,7 (5.89)	698.3	12.22	50.25	13.85
2	Small	19,753,2	1,600,0 (1.79)	902.7	16,49	19.54	17.91
3	Semi-medium	28,004,8	1,228,1 (0.91)	1,355,3	23,38	15.00	26.89
4	Medium	35,734,9	997,1 (0.78)	1,280.9	29.84	12.17	25.41
5	Large	21,639,9	248,9 (0.31)	802,5	18.07	3.04	15.92
	Total	119,765,4	8,934,5	5,125,9	100.00	100.00	100.00

Note : Figures in brackets are ratios of leased-in to leased-out area within the group.

As in the case of households, the ratio of leased-in area to leased-out area shows[5] an inverse relationship with the size of holding and the class of farmers. The marginal owners show a ratio of 5.89 while the corresponding ratio for the large farmers is 0.31. Medium and large farmers, as is to be expected, lease-out much more than what they lease-in. Despite significant leasing-in by large owners the dominating and persisting trend in this group is leasing-out. The leased-out area of the large farmers is nearly three times their leased-in area. Thus while capitalist tenancies are to be noticed the traditional patterns in the lease market persist.

Shares in Leased-In and Leased-Out and Class of Farmers

The share of leased-in area and leasing-in households by class of farmers shows an inverse relationship. More than a quarter of the households among the marginal and the landless households lease-in. Leased-in area forms around 30 per cent of the owned area of the marginal farmers. Among the small owners the proportions are less, but exceeds the corresponding proportion for medium and large groups of owners. Out of the 7.47 per cent of total owned area leased-in 3.75 per cent of owned area is accounted for by marginal owners. This may appear small absolutely. But its significance should not be under-rated. It is much higher than what is estimated to be ceiling surplus. Even if the leased-out area of the marginal owners is deducted, the net leased-in area of the marginal owners exceeds what is distributed under the ceiling-surplus.[6] Leased-in area of marginal owners is also not insignificant when related to their owned area which forms 12.22 per cent of the total owned area (Table 2.3).

TABLE 2.3 : Leased-in and Leased-out Households and Area: 1982

	Class of Farmer by Owner Group	*Households*		*Area*		*Percentage of Total Owned Area*	
		Leasing-in	*Leasing-out*	*Leasing-in*	*Leasing-out*	*Leased-in*	*Leased-out*
		(Percentage within group)					
1	Marginal including landless	26.19	4.11	28.13	4.77	3.75	0.59
2	Small	14.19	8.01	8.10	4.57	1.46	0.77
3	Semi-medium	11.23	9.46	4.38	4.83	1.12	1.15
4	Medium	8.22	6.16	2.79	3.58	0.91	1.09
5	Large	4.62	8.80	1.15	3.71	0.23	0.68
	Total	18.53	5.53	7.47	4.28	7.47	4.28

The weak positive relationship, both in terms of households and in terms of area, between the size of the holding and leasing-out could be more due to resumptions by medium and large farms, and under-reporting of leased-out area on large farms.[7] Even with possible under-reporting of the leased-out area, the net leased-out area on large farms is considerable.

At the all-India level, those who lease-out fully (pure rentiers) form only 7.41 per cent of the total rentiers. As is to be expected, among the class of marginal owners, the proportions are little higher than the average. Even so, marginal owners who improve their access to the labour market by completely leasing-out the land form less than 10 per cent (Table 2.4).[8] The remaining 90 per cent belong to the 'partly leasing-out' group. It is obvious that those who improve their access to land by leasing-in form several multiples of those who lease-out fully and seek to improve their access to the labour market. This suggest that for the landless and the marginal owners access to land market continues to be much more important even when they enter the labour market as rural labour.

TABLE 2.4: Distribution of Households and Area Leased-out Fully (In percentages to those Leasing-out Fully and Partly within the Group) All India Rural : 1982.

Class of Farmer (Hectares)	*Households Leasing-out Fully*	*Area Leased-out Fully*
Marginal :		
0.002-0.20	14.98	9.06
0.21-0.40	9.56	14.34
0.41-0.50	6.65	9.19
0.51-1.00	10.54	17.57
Small :		
1.01-2.02	2.12	3.28
Semi-medium :		
2.03-3.03	5.51	9.86
3.04-4.04	6.23	12.14
Medium :		
4.05-5.05	1.41	3.16
5.06-6.07	Nil	Nil
6.8-8.09	7.05	12.83
8.10-10.12	0.28	0.52
Large :		
10.13-12.14	Nil	Nil
12.15-20.24	Nil	Nil
20.25 and above	Nil	Nil
All	7.41	7.21

Regional Dimensions of Tenancy

When the states are arranged in the descending order of percentage area under tanancy,[9] which ranged between 19.65 in Haryana and 2 per cent in Gujarat, a pattern is seen to emerge (Table 2.5). The top tenancy states are comprised of two groups; high irrigation-intensive, high productive and technologically developed north-western states; and, densely populated, high rainfall, rice-growing and technologically less developed north-eastern states. Tamil Nadu is one state outside these two groups, but within the high tenancy group. It combines the high density of north-eastern states and the high irrigation intensity and technological development of north-western states. This group of states together accounts for around 11 million and more than 65 per cent of total tenants.

Among the high tenancy incidence states, Punjab and Haryana account for only 8.07 per cent of the total tenant households. The rest accounting for 57.75 per cent of the total tenant households in the country are concentrated in the states of Tamil Nadu, West Bengal, Uttar Pradesh, Bihar and Orissa all of which record higher than average incidence of rural poverty (Table 2.5).

Andhra Pradesh, Maharashtra and Karnataka have also pockets of high tenancy and have large number of tenants though they do not fall in the high tenancy category state-wise.

Terms of Leasing

At the all-India level, the distribution of leased-in area by terms of lease shifted from fixed money and fixed produce to sharecropping between 1960-61 and 1971-72. The data, for 1982 do not give a clear picture of trends since for 30 per cent of the area, records by terms of lease were not available. If one assumes that the non-reporting area is distributed in the same ratio as the reported, sharecropping should be considered to have risen between 1971-72 and 1982 (Table 2.6).[10]

State-wise, (Table 2.7) Bihar shows the highest concentration of area governed by sharecropping, followed by West Bengal and Uttar Pradesh. A similar, but not identical, picture emerges from the responses of those who lease-out (Table 2.8). A comparison of the two sets of figures throws interesting sidelights on the disparities between leased-out and lease-in area. One would expect a higher proportion of area under fixed money and fixed produce in the leased-in area, as compared to leased-out, since the urban households and non-household institutions are covered under the leased-in and not in the leased-out area and such households are generally known to prefer fixed rents.[11]

TABLE 2.5 : Distribution of Leased-in Households by States All-India; Rural :1982

State by Descending Order of Percentage Leased-in Area	*Percentage Households Leasing-in*	*Percentage Area Leased-in*	*Absolute number of Tenants (00s)*	*Absolute Tenanted Area (00h)*	*Percentage*		*Percentage Irrigated in Operated (1981-82)*
					Leased-in House-holds in total	*Leased-in Area in Total*	
1	2	3	4	5	6	7	8
High Tenancy States :							
1. Haryana	22.28	19.65	3735	5681	2.15	6.36	66.20
2. Punjab	23.03	18.98	5083	5871	2.92	6.58	78.00
3. Tamil Nadu	29.22	13.37	20574	5294	11.83	5.93	46.98
4. West Bengal	27.09	12.29	20908	5230	12.82	5.85	22.29
5. Uttar Pradesh	21.27	11.09	32781	17607	18.85	19.71	59.59
6. Bihar	17.22	10.42	19382	9284	11.14	10.39	24.82
7. Orissa	16.70	8.04	6784	3285	3.90	3.68	11.39
Medium Tenancy States :							
8. Assam	14.09	6.87	3180	1404	1.82	1.57	0.51
9. Karnataka	16.95	6.62	8168	5690	4.70	6.37	9.59
10. Andhra Pradesh	19.74	6.51	16956	7215	9.75	8.08	15.55
11. Maharashtra	16.70	5.57	12750	8390	7.33	9.39	10.12
Low Tenancy States :							
12. Rajasthan	9.70	4.34	4234	6430	2.43	7.19	11.62
13. Madhya Pradesh	12.09	3.80	8045	5376	4.63	6.01	10.75
14. Himachal Pradesh	16.95	2.94	937	202	0.54	1.05	9.88
15. Jammu and Kashmir	5.40	2.79	441	211	0.26	0.24	24.08
16. Kerala	12.70	2.28	4624	300	2.66	0.34	11.99
17. Gujarat	9.01	2.00	3514	1426	2.02	1.59	26.47
All-India (including States not specified)	18.53	7.46	173913	89345	100.00	100.00	24.84

Source : NSS 37th Round.

TABLE 2.6 : Percentage Distribution of Leased-in in Operated Area by Terms of Lease : All-India : 1981-82

NSS Round	*For Fixed Money*	*For Fixed Produce*	*For Share Product*	*For Share of Produce with Other Terms*	*Under Usufructuary Mortgage*	*Other Terms*	*Not Recorded*	*Total*
1	2	3	4	5	6	7	8	9
37th Round (1982)	10.86	6.27	41.92	2.79	2.23	5.15	30.78	100.00
26th Round (1971-72)	15.42	11.64	47.86	—	3.13	21.95	—	100.00
17th Round (1960-61 and 1961-62	25.61	12.90	38.22	0.93	—	22.34	—	100.00

Source : *Sarvekshana,* NSS Report 331, 37th Round, Table VII, p. 27.

TABLE 2.7 : Percentage Distribution of Leased-in in Operated Area by Terms of Lease : 1982 (In States with High Concentration of Tenancy)

State	*For Fixed Money*	*For Fixed Produce*	*For Share Produce*	*Share of Produce with Other Terms*	*Under Usufructuary Mortgage*	*On Other Terms*	*Not Recorded*	*Total*
1. Bihar	6.52	3.60	73.32	3.50	2.04	2.34	8.68	100.00
2. Haryana	24.15	10.81	41.22	5.27	0.38	4.88	13.29	100.00
3. Orissa	5.14	8.06	42.04	2.42	1.51	3.63	37.20	100.00
4. Punjab	42.13	4.60	39.89	3.92	2.05	3.98	3.43	100.00
5. Tamil Nadu	19.23	19.87	36.54	0.92	2.47	7.88	13.09	100.00
6. Uttar Pradesh	8.59	4.88	50.10	3.52	0.49	7.03	25.39	100.00
7. West Bengal	2.84	11.91	55.59	0.24	0.49	11.26	17.67	100.00
All-India	10.86	6.28	41.92	2.79	2.23	5.15	30.78	100.00

Source : *Sarvekshana*, NSS Report No. 331.

TABLE 2.8 : Distribution of Area (37th Round) Leased-out Under Different Terms in States with High Concentration of Tenancy: 1982

State	*For Fixed Money*	*For Fixed Produce*	*For Share Produce*	*Share of Produce with Other Terms*	*Under Usufructuary Mortgage*	*On Other Terms*	*Total*
1. Bihar	14.55	16.36	57.17	0.81	9.70	1.41	100.00
2. Haryana	28.20	33.27	29.14	2.07	2.35	5.08	100.00
3. Orissa	24.22	18.53	40.55	0.00	2.57	13.94	100.00
4. Punjab	34.24	11.20	30.98	3.88	10.39	9.39	100.00
5. Tamil Nadu	30.90	5.60	34.28	1.36	22.41	7.30	100.00
6. Uttar Pradesh	3.97	2.09	65.34	11.69	1.87	15.24	100.00
7. West Bengal	5.65	19.76	53.22	8.47	2.02	10.48	100.00
All-India	20.51	12.59	45.92	3.03	4.43	13.52	100.00

Source : *Sarvekshana*, October 1987.

Surprisingly the opposite is the case. Most likely sharecropping is under reported by owners when they reported leased-out. the percentage of sharecropped area, when adjusted for 'non-reported area' is much larger in the leased-in data as shown in Table 2.7.

It is only in the north-western states of Punjab and Haryana, in which the percentage of areas leased-in by the landless and marginal forms a small proportion, and in Tamil Nadu the area governed by leases fixed in money and produce is relatively more significant. In all other states sharecropping continues to be not only significant but appears to be rising in its incidence.

Sharecropping coupled with short-term leases, it is sometimes maintained, ensures productive efficiency, which is no less than other terms of lease (Cheung, 1969). But when it is associated with rack-renting, labour rent (Murthy, 1987) and self-exploitation of the tenant households it affects human resource development. Recent studies by ICRISAT scholars (Ryan and Walker, 1990) cast doubts also on the productive efficiency of sharecropping under conditions of shuffling of tenants and short-term leases. In their study villages in the dry zone sharecropping accounted for 18.7 per cent reduction in output and a sizeable decline in the average use of family labour and bullock draft. Farmers were found to sow more extensive, lower-valued crops on sharecropped land (Walker and Ryan, pp 178-179).

In brief, the terms of lease do not appear to have shifted away from share-cropping, and sharecropping with short leases and generally with no cost-sharing, could not only be exploitative, but affect productive efficiency.

Technological Development, Leasing-in and Leasing-out : Punjab and Bihar

Technological development is said to contribute to two changes in the lease market: first, emergence of capitalist tenancy; and second, entry of the small owner, who seeks to improve his position in the labour market, into the land market as a rentier. The first change reduces the access of the landless, marginal and small to land, and the second suggests declining importance of access to land for the small. We examine these by a study of the comparative position of Punjab and Bihar (Tables 2.9 and 2.10).

The most distinctive aspect of the lease market of Punjab as compared to Bihar is the entry of the middle farmer group as leasers-in.[12] While in Bihar leasing-in households show a distinct inverse relationship with the size of holding, the relationship in Punjab is of an inverted 'U' shape, rising as we proceed from marginal to semi-medium farming group and then falling as we move from medium to large. There is capitalist tenancy as distinguished from subsistence tenancy no doubt, but it is much stronger in the middle groups as compared to large farmer group. The contrast is less pronounced in respect of 'leasing-out' pattern. In Punjab, as in Bihar, the percentage of leasing-out households rises with the size of holding except for the interruption of the pattern by the middle group, which shows less percentage of households leasing out as compared to the semi- medium. The emergence of the middle farmer as a capitalist tenant does restrict access to land of the marginal and small farmer. But these trends continue to be associated with the persistence of considerable leasing-out by the large farmer group. Nearly a third of the total households lease-out in this group.

TABLE 2.9 : Households Leasing-in and Leasing-out (Percentages in each Group Bihar and Punjab : 1982

Size Group of Ownership (in Hectares)	*Leasing-in*		*Leasing-out*		*Net Leasing-in Over Leasing-out*	
	Bihar	*Punjab*	*Bihar*	*Punjab*	*Bihar*	*Punjab*
Marginal and Landless						
1 Below 1.00	19.27	17.15	5.69	5.53	13.58	11.62
Small:						
2 1.01-2.02	14.14	38.10	12.24	9.64	1.90	28.45
Semi-medium:						
3 2.02-3.03	7.21	40.60	11.25	17.67	–4.04	29.39
4 3.04-4.04						
Medium:						
5 4.05-5.05						
6 5.06-6.07						
7 6.08-8.09	4.22	25.62	14.98	11.85	–10.76	13.77
8 8.10-10.12						
Large:						
9 10.13-12.14						
10 12.15-20.24	3.03	18.41	23.63	30.36	–20.60	–11.95
11 20.25 and above	17.22	23.03	7.44	8.35	9.78	14.68
All size						

As regards leasing-out by the marginal and the small farmer, this is as yet restricted to a small proportion of the owners. What is to be noted is

TABLE 2.10 : Distribution of Households, Owning, Leasing-in and Leasing-out in Total Bihar and Punjab: 1982

Class of Farmer	*Owning*		*Leasing-in*		*Ratio of Leasing-in to Leasing-out Households*	
	Bihar	*Punjab*	*Bihar*	*Punjab*	*Bihar*	*Punjab*
1	2	3	4	5	6	7
1. Marginal (inclu ding landless)	76.65	66.87	85.79	50.21	3.39	3.10
2. Small	12.42	10.08	10.20	16.81	1.16	3.95
3. Semi-medium	7.79	11.61	3.26	20.64	0.64	2.30
4. Medium	2.82	9.94	0.69	11.14	0.28	2.16
5. Large	0.31	1.47	0.06	1.19	0.13	0.61
Total	100.00	100.00	100.00	100.00	—	—

that the percentage of owners leasing-out among marginal and small farmer in Punjab is not higher than what we notice for Bihar. Even the ratios of those who lease-in to lease-out among marginal owner and small owner groups in Punjab do not differ much from those observed for Bihar. Even in Punjab a significant proportion of the marginal farmers, and a much larger proportion of the small farmers continue to lease-in. Technological development, while it has contributed to the emergence of capitalist tenancy, has not reduced the importance of subsistence tenancy even in Punjab. Around on-half of the area leased-in in Punjab continues to be accounted for by the marginal and small farmers, and such area forms 57 per cent of the owned area of these groups.

As the labour market gets tightened, we should expect marginal owners to shift from the lease-market to the labour market increasingly. If there is no such trend we may infer that the access to land continues to be important for the marginal owners. We have identified two group of marginal owners which account for the bulk of the marginal owners in all the states. In both the groups area leased-out as a percentage of the owned area showed a decline in all the states except in Tamil Nadu. Even north-western states of Punjab and Haryana show a general trend of decline (Table 2.11).

Poverty Alleviation and Policy Options for Tenancy Reform

Analysis shows that the decline in tenancy reduced the overall access to land of marginal owners and landless. Despite some trends of reverse leasing, which were seen in commercialised areas even in the

TABLE 2.11 : Leased-out Area as Percentage of Owned Area (Among Marginal Owner Groups) in States with High Concentration of Tenancy : 1971-72 and 1982

State *Size Group*	*0.002-0.20*		*0.51-1.00*	
	1971-72	*1982*	*1971-72*	*1982*
1. Bihar	12.22	4.09	6.85	4.32
2. Haryana	11.83	0.19	14.82	9.03
3. Orissa	12.72	3.77	10.26	5.52
4. Punjab	4.54	4.61	29.76	26.78
5. Tamil Nadu	5.38	6.32	4.12	7.58
6. Uttar Pradesh	9.83	4.14	8.91	5.63
7. West Bengal	5.59	1.00	7.78	2.19

Source : *Sarvekshana*, Vol V, Nos. 3 and 4, January-April, 1982 for 1971-72, and Vol. XI, No. 2, Issue No. 33, October, 1987 for 1982.

pretechnology period, the basic aspect of tenancy continues to be subsistence tenancy. Further, the bulk of small tenancy is concentrated in the states with relatively high incidence of poverty. In these areas, renting out by the small owners is not only insignificant, but over a period shows signs of decline, suggesting the continued importance of access to land for landless and marginal owners. Moreover, in areas with high concentration of tenancy sharecropping is not only significant but is on the rise. Therefore, an effectively implemented tenancy legislation has not only the potential for poverty alleviation, but could contribute to growth.

A rough estimate of the possible benefits of improved access to land through tenancy reform shows that a fourth of the marginal owners including the landless could benefit. Even if the tenants among marginal owners, many of them agricultural labourers are estimated to be 15 per cent of the total rural resident households, they could be around 30 per cent of the poor households. These could gain by effective tenancy legislation ensuring security and effective regulation of rents at the legislated level of 20 to 25 per cent of the gross produce as against the market (black market) levels of 50 per cent. For the small owners, who lease in on the average 0.82 h the gains could be substantial and reforms could lift them above the poverty line. This is on the assumption that the gross value of output per hectare is Rs. 3976 (figure obtained from the National Account Statistics for triennium ending 1982-83 at 1980-81 prices), and the gain in terms of rent at 25 per cent instead of 50 per cent is Rs. 815. The gain in income as a percentage of poverty line of Rs. 6,000 per household is 13.58. On similar assumptions relating to gross value and rental gain, the marginal owners (with an average rented area of 0.25 h) would gain only Rs. 244 per annum.[13] This indicates that tenancy reform should be seen as a means to provide an asset base for pursuing land-based and other supplementary income-generating activities and also as a means for widening the access to improved technology and credit. But in double cropped irrigated areas, as in the deltas of Godavari and Krishna of Andhra Pradesh, even for leased in area of one-fourth of a hectare, the marginal owners could have a saving of 8 bags of paddy of 600 kgs costing Rs. 1,200 per annum.[14] The gains to the tenant through effective implementation of tenancy reform in such areas could be much more than what any DRDA programme could offer.

Policy Options

There could be no doubt about the relevance of tenancy reforms in poverty alleviation especially in areas with high concentration of tenancy. The most important question then, is whether such reforms should be pursued at all. Gunnar Myrdal (1968) preferred no reforms to ineffective reforms resulting in less access and more insecurity to small tenants Daniel Thorner (1978) preferred no action to limited and timid action. He wrote:

> "If India's agrarian history demonstrates any thing, it is that doing and saying nothing is preferable to taking small steps timidly and slowly."

The World Developed Report of the World Bank (1990) preferred policies that raise the demand for unskilled labour and boost the returns to labour to tenancy laws which are not implemented. Further arguments also could be advanced in favour of no action in the area of tenancy reform. Competitive forces would bring down the level of rents as a share of produce due to (1) tightening of the labour market; (2) increased productivity per hectare; and (3) shift in the system of rental from share to fixed system. Moreover, in areas, in which there could be a profitable shift of labour from agriculture into non-agriculture, rigidities in the land market will obstruct mobility of labour and will be counter-productive. It is further argued that the land reform of the 1950 and 1960 and the technology of the 1960 have curbed feudalism within Indian agriculture and lead to the dominance of the owner-cultivating middle peasant entrepreneur, and the remnants of tenancy that we see have a legitimate function.

The arguments for not doing any thing with tenancy are not as powerful as they are made to appear. The opportunities for absorptions of marginal owners and landless labour in non-agricultural labour are not rising fast enough to tighten the labour market. In states with high degree of concentration of tenants, the bulk of the area is under subsistence tenancy and the bulk of land leased out is by medium and large farmers whose numbers are small in each village. Further a significant proportion of area leased-out is by urbanities who have non-agricultural incomes. 50 per cent or more of the area is governed by share cropping arrangements under which all costs continue to be borne by tenants. More recent studies in the villages of East Godavari and West Godavari delta show shifts to single season leasing to landless labour by big resident landlords who are now engaged increasingly in non-agricultural pursuits. There is re-emergence and growth of insecure tenancy in the irrigated areas. Hence at least in areas in which 'subsistence tenancy' continues to be prominent, though not recorded, registration of tenants would guarantee their security, assist them in obtaining loans from institutional agencies, and increase the certainty that they would benefit from increased production. Therefore, the suggested policy option is a vigorous pursuit of registration of informal tenants and protection of their existing rights.

But, guarantee of security of tenants through registration of their record of rights proved to be impracticable when owners preferred otherwise in the absence of measures followed in successful models of tenancy reform such as in Kerala and in West Bengal. The success of the Kerala model is attributed to several factors. Invoking Section 72 of the Kerala Land Relations Act, Government of Kerala ordered that on January 1, 1970, the entire rights, titles and interests of holdings held by the cultivating, tenants vested in the government and they became owners of their leased-in holdings. Thus, landlordism was abolished as if by the stroke of a pen.[15] Resumption for personal cultivation is permitted in the Kerala model only in the case of small farmers, i.e. those owning less than two hectares and that too to make it up to that size under severe restrictive conditions, so that the clause ceases to be the permissive loophole (used for eviction and even rack-renting) which one finds in similar legislations in other states in the country. The implementation of the abolition of landlordism is almost complete today. The thoroughness and speed with which tenancy abolition has been carried out in Kerala has no parallel elsewhere in India. But the major factor for the success of land reforms in Kerala is that 'land reforms of Kerala have been the culmination of long drawn-out struggle that goes back to the preindependence days. There was mass mobilisation of the peasants and agriculture workers under the CPI banner throughout Kerala. Oomen sums up the lesson of Kerala model:

> "I do not propose of elaborate except to maintain that unless land reforms are backed up by mass mobilisation with a government sympathetic to the working class both at the states and the centre, they cannot succeed. This is an important lesson to be drawn from the experience of Kerala" (Oomen, 1990 p. 31)

The West Bengal model differs from Kerala model in important respects. In Kerala, the entire rights, titles and interests of holdings held by the cultivating tenants are vested in the government and they became owners of their leased-in holdings, of course, subject to the ceiling provisions. In West Bengal, tenants have not been transformed into owners, but have been assured protected tenancy and fair rents.

In West Bengal three concrete measures were taken to make it easier for the tenants to prove their status. Firstly, the onus for proving that a person, cultivating the land was not a tenant was put on the landowner—it was now for him to bother with lawyers, court and legal expenses. Secondly, the landowners were required to issue a receipt in exchange for crop-share received. The third and most popular measure was to launch a campaign for recording the sharecroppers (*bargadara*) code-named 'Operation Barga', which would speed up the established, but time-consuming, method of recording through the settlement machinery. The procedure adopted is described as follows :

> The first step in the campaign was to identify areas with an expected high level of concentration of sharecropping; to set-up camps there, where a group of officials would meet villagers collectively. The collective strength of the officials would counter the possibility of intimidation by the vested interest, while the collective participation of the villagers in the camp would make

> it easier to verify the claim of sharecroppers on the spot. . . . An important aspect of this campaign was to make the intended beneficiaries aware of their rights. It was during this campaign that the vast majority of sharecroppers came to know of the legal rights which had remained unknown to them and left unimplemented for more than two decades. (Biplab Das Gupta, 1982 p. 15)

Besides the 'Operation Barga', there were stringent provisions relating to resumptions for self-cultivation.

> Firstly, to resume such land for self-cultivation, the landowner would have to prove that he or members of his family line are near that land during most of the year. Secondly, he should also prove that most of his income is derived from such land; those having other sources of substantial income would not be given any opportunity to take over land from the sharecropper on this ground. Thirdly, 'personal' cultivation would mean exactly this: that is, cultivation by himself or by members of the family directly, and not by way of employment of farm servants or hired labourers. Lastly, such resumption of land by the owner should leave a minimum of 2.5 acres with the sharecropper for continued cultivation. (*Ibid*, p. 16)

The major implications of the experience in West Bengal are as follows:

> First, the 'political will' of the government , while the most crucial factor, would not by itself bring about a radical change in land relations without organised mobilisation and active participation of the intended beneficiaries in the programme. (*Ibid*, p. 18)

In both the models, there are two common elements: (1) political will for successful implementation of the reform, and (2) successful mobilization of the beneficiaries in favour of the reform. There is need to review tenancy legislations in other states in the light of these two successful models, follow approaches similar to 'Operation Barga' of West Bengal, established quickly with a time target record of rights for all informal tenancies of marginal owners and the landless, by starting with areas with high concentration of tenancies and with rented land of the non-resident landowners and medium and large resident owners leasing out lands.

If the renewed rhetoric on land reform is not supported by required political will and mobilisation of support for the reform at grassroots level,[16] the most likely prospect is further deterioration of access to land for marginal owners and landless, growth in informal tenancy and more insecure conditions for tenants affecting not only equity but also growth.

Note: Reproduced from, *Economic and Political Weekly*, March 30, 1991 with kind permission of the publisher.

NOTES

1. The Bombay Tenancy Act of 1956, for the first time in India implemented a land-to-the-tiller law. As a result there was very large-scale transfer of ownership of leased land to tenants.
2. For instance see Herring (1983 p. 57) in support of the contention that the act benefited mainly the upper class of tenants.
3. What is maintained here is that reverse tenancy is not a new phenomenon but a growing phenomenon. Rath drew my attention to the publications of Dandekar in *Artha Vijnana* in 1958-59 and Rath in 1962.
4. The ICRISAT longitudinal study (Walker and Ryan, 1990 p. 17) comments: "It was only after two years of observing respondents working in their land and others' fields and others working in respondents' fields and cross checking information in the plot cultivation, resource endowment and transaction schedules that we were reasonably confident that the tenancy data were reliable."
5. The total leased-out area recorded is less than the total leased-in area. The excess of leased-in area is probable insofar as leased-in area included area from urban households and non-household institutions while leased-out are details elicited from the rural households and do not include these. The two sets of figures do not match even when we consider the figures shown under 'urban households' and 'non-household institutions' in both sets of figures. There is an obvious under-estimation of leased-out area most possibly in the upper group category of farmers. If under-reporting is considered the observed relationship will be much stronger.
6. As against the estimated availability of surplus land ranging between 120 lakh acres to 302 lakh acres computed by various surveys, the total area declared surplus so far has been 73.35 lakh acres only (less than 2 per cent

of the cultivated area) of which 44.45 lakh acres has been distributed among 41.19 lakh beneficiaries.
From background note paper prepared by the Planning Commission circulated for the seminar on 'Land Reforms—Retrospective and Prospect' held on June 7 and 8, 1989.

7. It is reasonable to think that under-reporting of both owned and leased-out area could be more in bigger size groups, who are likely to be affected by land reforms, than under smaller size groups.
8. Rath drew my attention to the fact that the definition of owned land includes all non-cultivated land (including house site) and therefore it may not be correct to think that the 90 per cent of marginal landowners who partly lease-out are partly cultivating. This may be the case with tiny among marginal owners but even within the group owning 0.51 to 1.00 h only around 10 per cent fully lease-out. The 90 per cent who do not fully lease out cannot possibly be holding their land as house sites.
9. The classification used is in relation to the average percentge of area under tenancy, low, below 7 per cent; medium, 7-14 per cent; and high, 14 per cent and above. There are other states in which there are pockets of high tenancy though the overall incidence at the stage-level is medium or low. Tenancy proportions may be much larger than reported in the NSS data, though the data are got from households leasing-in.
10. Swamy in the *IJAE* article draws the inference that sharecropping declined. The inference is drawn from data relating to 'lease-out' which is not only partial but also less reliable than data on leased-in.
11. Rath has pointedout in his communication to me that urban households could also be leasing out on share-term basis. But the issue is disparity in the proportions of area under each contractual term in leasing-in and leasing-out data. We place more reliance on leasing in data. Further there are strong reasons to believe that when the owners report the leased-out area. They are most likely to suppress the sharecropped area.
12. This is not however a new phenomenon. Even in the late fifties, long before the new technology, in the more developed districts higher proportions of rented-in area were noted on the farms of big occupations (Rastiannikov, 1981 pp. 242-47).
13. Region-specific data will be useful. But our purpose here is to indicate the broad dimensions of benefits at the aggregate level.
14. The fixed rent per season is 12 bags of paddy, on land producing 24 bags per acre or 60 bags per hectare, amounting to 50 per cent of the gross produce which the legislated rent is 25 per cent. For two crops of paddy the rent paid for 0.25 hectare is 16 bags while the legislated rent is 8 bags (Bag: 75 kgs).
15. The Bombay Tenancy Act, 1948 and amended in 1955, did provide for compulsory vesting of the landlords rights, in tenants, but only for

'Protected tenants' after the weakest sector of the tenancy had been dispossessed (Herring, 1983 p. 157).

16. Chadha and Bhaumik point out that the organisational base of agricultural labour has been strengthened only in recent years with the revival of panchayat raj system. (Chadha, G K and Bhaumik, S K, 'The Changing Tenancy Relations in West Bengal, Popular Nations, Grassroots Realities-II, forthcoming in *EPW*).

REFERENCES

Appu, P. S. (1975) : 'Tenancy Reform in India', *EPW*, Vol X, Nos 3335, Special Number, August, pp 1353-55.

Bhalla, Sheila (1983): 'Tenancy Today : New Factors in Determination of Mode of Level of Payments in Agricultural Land', *EPW*, Annual No May.

Cheung, S. N. S. (1969): *The Theory of Share Tenancy*, Chicago University Press, Chicago.

Dasgupta, Biplab (1982): 'Some Aspects of Land Reform in West Bengal', *Land Reform*, 982, No 1/2, FAO, Rome, pp 12-20.

Gill, S. S. (1989):'Changing Land Relations in Punjab and Implications for Land Reform', *EPW*, June 24.

Herring, Ronald J., *Land to the Tiller*, Delhi, Oxford University Press, p 212.

Khusro, A. M. (1958): *Economic and Social Effects of Jagirdari Abolition and Land Reforms in Hyderabad*, Hyderabad, Osmania University Press..

Murthy, K. S. (1987): 'Impact of Technological Change on the Extent of Tenancy and on Composition of Tenants: A Study in Two Villages', *IJAE*, July-September, pp 286-293.

Myrdal, Gunnar (1968): *Asian Drama: An Enquiry into the Poverty of Nations*, New York, Random House.

Oomen, M. A. (1990): 'Land Reform and Economic Change: Experience and Lessons from Kerala', *Science and People*, Vol 1, No 2, March, pp 28-38.

Rajagopalan, R. and Anuradha, B. (1987): 'Structural Changes in Land Holdings of Thanjavur District of Tamilnadu', *IJAE*, Vol XVII, No 3, Conference Number, July-September, pp 307-312.

Rastiannikov, V. G. (1981): *Agrarain Evolution in a Multiform Structure Society: Experience of Independent India*, Routledge and Kegan Paul, London.

Singh, Iqbal (1989): 'Reverse Tenancy in Punjab Agriculture: Impact of Technological Change', *EPW*, June 24.

Swamy Dalip, S. (1988): 'Agricultural Tenancy in the 1980s', *IJAE*, October-December, Vol XLIII, No 4, pp 555-68.

Thorner, Daniel (1973): *The Agrarian Prospect in India*, 2nd Ed, Columbia University, South Asia Books.

Walker, S. Thomas and Ryan James, G. (1990): *Village and Household Economies in India's Semi-Arid Tropics*, The Johns Hopkins University. Press; Baltimore and London, pp 170-79.

World Bank (1990): *World Development Report*, 1990, Oxford Univeristy Press.

3

CHANGING LAND RELATIONS OF PUNJAB AND IMPLICATIONS FOR LAND REFORMS

SUCHA SINGH GILL

Land relations is an important factor having implications for production, productivity and distribution of income and wealth in agriculture. Proper handling of the land question is a serious policy issues for the state. After independence the Congress Agrarian Reforms Committee report brought out the significance of this issue. On the basis of this report various state governments started the process of agrarian reforms legislation. The intention was to bring !and relations in tune with the requirements of capitalist development in agriculture. Once this purpose is achieved the relevance of the earlier legislation assumes a changed context. Here our purpose is to examine the changing land relations and relevance of the existing land reform measures in Punjab. There are three sections:

Section I deals with land relations on the eve of introduction of land reforms. Section II deals with land reform measures and their implementation and the last section covers recent changes in land relations and their implications for the existing land reform measures.

I

At the time of partition of the country, a major part of the land was under tenant cultivation. Land revenue records of Punjab indicate that land under tenant cultivation was 48.6 per cent in 1947 and 48.2 per cent in 1952[1]. This is also confirmed by a field report of three villages in Punjab. According to this report the area under tenant cultivation was between 37 per cent and 50 per cent in 195[2]. The eighth round of National Sample Survey estimated that in 1953-54 the extent of land under tenant cultivation was 40.42 per cent in Punjab (including Delhi and (Himachal Pradesh) and 37.71 per cent in PEPSU[3]. The Farm Management Survey data related to Ferozepur district show that the percentage of operated area under tenancy was 37 in 1954-55[4]. It is evident that at the time of enacting of land reforms legislation in Punjab and PEPSU nearly 40 per cent of the operated area was under tenancy. Of the total leased out area 49.52 per cent in Punjab and 32.83 per cent in PEPSU was leased out by non-cultivating landowners. The remaining area was leased out by cultivating owners[5].

Tenancy was of two kinds : (a) occupancy, and (b) at will. In Punjab, occupancy tenants with inheritable rights cultivated 7 to 10 per cent of the cultivated area and tenants-at-will cultivated between 30 and 40 per cent of the area in 1952[6]. According to the report of PEPSU government in 1952 there was 6,16,799 acres under occupancy tenants and 10,46,532 acres under tenants-at-will[7]. Thus the major part of tenanted area was under tenanta-at-will with no security of tenure.

The tenants were further of two types: (a) purely tenants, and (b) owner-cum-tenants. Purely tenants did not own any land whereas owner-cum-tenants leased in land to increase the size of their holdings. Tables 3.1 and 3.2 show the percentage distribution of land leased out and leased in. In Punjab 1951 per cent and in PEPSU 16.90 per cent of the households were that of purely tenants and owners-cum-tenants

TABLE 3.1 : Percentage Distribution of Area Leased Out for 1950-51 to 1953-54

	Livelihood Class	*Percentage of Estimated Households*	*Percentage of Area Leased out*			
			1950-51	*1951-52*	*1952-53*	*1953-54*
		Punjab				
1.	Cultivating owners	35.59	37.08	35.64	36.45	38.10
2.	Non-owning cultivators	19.51	5.65	5.63	5.49	6.97
3.	Non-cultivating owners	3.54	50.67	52.07	50.62	49.52
4.	Others	40.96	6.60	6.66	7.44	5.41
	Total	100.00	100.00	100.00	100.00	100.00
		PEPSU				
1.	Cultivating owners	35.85	52.93	53.93	53.65	50.38
2.	Non-owning cultivators	16.90	2.92	2.77	2.79	3.14
3.	Non-cultivating owners	7.27	30.77	30.35	30.53	35.88
4.	Others	39.98	14.08	12.95	13.03	13.60
	Total	100.00	100.00	100.00	100.00	100.00

Source : NSS, 8th Round, No. 59. p. 37.

TABLE 3.2 : Percentage Distribution of Area Leased in for 1950-51 to 1953-54

	Livelihood Class	*Percentage of Estimated Households*	*Percentage of Leased in Area*			
			1950-51	*1951-52*	*1952-53*	*1953-54*
		Punjab				
1.	Cultivating owners	35.99	21.87	23.56	27.13	23.07
2.	Non-owning cultivators	19.51	65.00	63.86	61.43	63.70
3.	Non-cultivating owners	3.54	0.28	0.27	0.18	0.18
4.	Others	40.95	12.85	12.31	11.26	8.05
	Total	100.00	100.00	100.00	100.00	100.00
		PEPSU				
1.	Cultivating owners	35.85	57.52	53.93	53.65	50.38
2.	Non-owning cultivators	16.90	36.98	2.77	2.79	3.14
3.	Non-cultivating owners	7.27	—	—	0.71	0.58
4.	Others	39.98	5.50	6.97	7.00	5.51
	Total	100.00	100.00	100.00	100.00	100.00

Source: NSS, 8th Round.

were 35.99 and 35.85 per cent respectively in Punjab and PEPSU area. They were leasing in land from both non-cultivating owners and cultivating owners. In Punjab non-cultivating owners' contribution was the largest in leasing out land and accounted for nearly 50 per cent, followed by cultivating owners whose contribution varied between 35.64 per cent and 38.10 per cent of the total leased out area.

Most of the land under tenancy was without any formal contract. Informal contracts accounted for more than 70 per cent of the land. Sharecropping was the most important method of payment of rent. Under this system, landlord and tenant shared crop on 50:50 basis. Besides, tenants were subjected to render services of various kinds to landlords. The distribution of land leased in under different types of contracts is shown in Table 3.3

Land, which is a major source of earning and one's position in the rural area, was concentrated in a few hands. In Punjab 35.51 per cent and in PEPSU 41.52 per cent of the households were without land. The farmers owning 5 acres or less constituted 35.04 per cent in Punjab and 20.82 per cent in PEPSU of the total rural households. The share of land was respectively 10,80 per cent and 5.46 per cent in Punjab and PEPSU. At the top, farmers who owned more than 30 acres of the land were 2.74 per cent in Punjab and 3.91 per cent in PEPSU. Their share of land was respectively 30.60 per cent and 28.03 per cent. This is evident from Table 3.4.

TABLE 3.3 : Percentage Distribution of Area Leased in for 1950-51 to 1953-54

Livelihood Class	*Percentage of Area Leased out*			
	1950-51	*1951-52*	*1952-53*	*1953-54*
	Punjab			
1. With formal contracts:				
1. Cash rent	3.60	3.28	4.85	6.95
2. Fixed crop rent	085	0.82	1.09	1.31
3. Proportionate crop rent	18.24	20.05	18.47	17.99
2. Without formal contracts:				
1. Cash rent	10.81	10.82	9.95	5.26
2. Fixed crop rent	2.31	3.04	2.71	3.04
3. Proportionate crop rent	27.75	26.75	31.42	37.10
3. Other types	14.14	13.67	12.75	13.17
4. Unspecified	22.30	21.57	18.76	15.18
Total	100.00	100.00	100.00	100.00
	PEPSU			
1. With formal contracts:				
1. Cash rent	24.37	17.69	18.56	15.77
2. Fixed crop rent	—	—	—	—
3. Proportionate crop rent	0.91	0.93	5.43	11.16
2. Without formal contracts:				
1. Cash rent	—	1.80	0.01	2.07
2. Fixed crop rent	—	—	—	0.03
3. Proportionate crop rent	10.64	13.20	14.55	11.82
3. Other types	15.49	15.62	18.69	16.52
4. Unspecified	48.59	50.76	42.76	42.63
Total	100.00	100.00	100.00	100.00

Source : NSS 8th Round.

TABLE 3.4 : Percentage of Households Owned Land Below Specified Size of Household Ownership Holding and Cumulative Percentage of Area Owned by Them for Major Crops Season 1953-54 in Punjab and PEPSU

Size of Household Ownership (in acres)	*Punjab*		*PUPSU*	
	Household	*Area Owned*	*Household*	*Area Owned*
1. 0-0.0005*	35.51	—	41.82	—
2. 1.00	50.80	0.65	50.54	0.39
3. 2.50	60.34	3.68	56.77	2.02
4. 5.00	70.55	10.80	62.64	5.46
5. 10.00	83.92	28.39	77.23	21.84
6. 20.00	93.66	53.40	90.40	50.96
7. 30.00	97.26	69.40	96.09	71.97
8. 50.00	98.99	81.23	99.11	90.30
9. 50 and above	100.00	100.00	100.00	100.00

Note : *Household either owning no land or land less than 0.005.

Sources : NSS, 8th Round.

The holdings of various sixes were faced with problems of fragmentation. This problem was highlighted by some studies as early as in 1920s. A study of village Bairampur showed that 28 per cent of the holdings had over thirty fields and in three cases plots were so small that they had been lost sight of by their owners and passed into the hands of others[8]. In the village Suner (Ferozepur) it was found that 50.3 per cent of the holdings had one to five fragments and 18 per cent had 6 to 10 fragments and remaining had fragments between 11 and 20[9,10]. Though, in 1930s there was some voluntary consolidation of holdings the problem of fragmentation remained unsolved till compulsory consolidation of holdings was introduced after independence.

The existence of intermediaries like biswedars, talukdars, ect., made the situation worse. They had no direct connections with land cultivation. But they exploited cultivators in various ways and forced begar (free labour) on them.

II

Land reforms adopted after independence in the state were intended to transform the land relations to suit capitalist modernisation. They can be divided into four categories.

1. Abolition of intermediaries;
2. Tenancy reforms;
3. Ceiling on landholding; and
4. Consolidation of holdings.

Toabolish intermediaries, the Punjab Abolition of Ala Malikiyat and Talukdar Act 1951 was enacted as President's Act. This was re-enacted in 1952. In October 1952, the institution of Zaidari was abolished and in the budget session of the state legislature, in 1953, a bill was passed for the abolition of the system of *Jagirdari* in the state. In PEPSU area PEPSU Abolition of Biswedai Ordinance was issued in 1950.

In both the areas occupancy acts were passed vesting proprietary rights in the occupancy tenants. The final shape to tenancy legislation was given in Punjab Securing of Land Tenures Act 1953 which was amended in 1955 and PEPSU Tenancy and Agricultural Lands Act 1955.

In both Punjab and PEPSU Acts, the maximum rent was fixed as 1/3rd of the gross produce. With these acts occupancy tenants for the ownership rights. But landowners had the right to eject tenants-at-will and take over land upto 30 standard acres (up to 40 standard acre in case of displaced person in PEPSU and 50 standard acres in Punjab). After ejecting tenants landowner could either cultivate the land himself or could give it to another tenant. The laws provided that a tenant was not to be ejected from an area up to five acres (including any other land held by him) until he had been settled on alternative land. In PEPSU laws security given to the tenants was upto 5 acres of land whether a tenant held land from big or small owner. In Punjab laws, a tenant on big holdings was not to be ejected at all until alternative land had been provided while a tenant on small holdings had been given security up to 5 standard acres. PEPSU laws gave security of tenure up to 15 standard acres to a tenant with continuous possession of 12 years at the commencement of the act (December 3, 1953).

Both Punjab and PEPSU laws contained provisions enabling tenants to voluntarily acquire ownership in the non-resumable area if they had continuous standing of 6 years in Punjab provision was 3/4 th of the market value and in PEPSU 90 times the land revenue (including rates and cess) or Rs. 200 per acre, whichever was less.

In PEPSU ceiling was fixed at 30 standard acres (in case of displaced persons 40 acres). In Punjab the state government was empowered to take over surplus lad above the permissible limit of 30 standard acres (up to 50 acres in case of displaced persons).

In Punjab (both areas) exemptions were granted to land in co-operative garden colonies, land granted for gallantry to army personnel; orchards, well-run farms and tea estates.

Following the guidelines of the Central and Land Reform Committee announced on August 3, 1971, Punjab legislature enacted Punjab Land Reform Act, 1972. A new level of ceiling was fixed in punjab (including PEPSU areas) at 7 hectares for double cropping land under assured irrigation, 11 hectares for one crop land under assured irrigation, 20.5 hectares of barani land and for other types of land at 21.8 hectares.

Ceiling had been applied to family as a unit rather than to individual members. In Punjab Act (1972), the above mentioned ceiling applies to family of five. When the number of members of family exceeds five, the permissible area increases by one-fifth of the permissible area for each member in excess of five, subject to the condition that additional land shall be allowed for not more than three such members.

Exemption had been given to land belonging to any religious or charitable institutions but not belonging to or vested in or taken in lease by the state, Central Governments, university and educational institutions or trusts, corporations owned or controlled by state and institutions or trusts, corporations owned or controlled by state and Central Governments, Bhoodan Yagna Board, agricultural co-operative societies, land mortgage banks and state or central co-operative banks[11].

In both Punjab and PEPSU, laws were enacted for compulsory consolidation of holdings to solve the problem of framentation. A separate department was opened by the state for this purpose and immediately work on consolidation was started.

The land reform legislation of Punjab (and PEPSU) had many flaws in it. The passing of land reforms legislation took a relatively longer time. A large number of landowners ejected tenants on paper and as well as in practice immediately after independence when the land reforms became a burning question. The situation regarding ejectment as reported by the Punjab government in the state assembly is given in Table 3.5.

TABLE 3.5 : Ejectment of Tenants by Landlords

Year	*Number of Ejectment Noticed*	*Number of Actual Ejectmentr*
1950-51	8388	3137
1951-52	7083	3287
1952-53	17503	NA
1953-54	60358	13700

Source : B.D. Talib, Agrarian Tensions and Peasant Movement in Punjab (Mimeo), CRD, JNU, New Delhi, 1979.

These figures are only about the Punjab area. The situation in PEPSU area, though not actually know, was not different. In addition to this a large number of cases of ejection were not allowed by powerful landowners to come to the notice of the state. The landlord interest dominated the state assemblies and wielded strong influence in police and local administration. They in collaboration with administration were able to suppress facts about the number of actual ejectments[12]. This resulted in reduction of area under tenancy. This is born out by the survey of four villages of Punjab and one of PEPSU by the committee on tenancy reforms in March 1956. As compared to nearly 40 per cent of area under tenancy in 1953-54, the committee found that in 1956 the area under tenancy was 30 per cent in Punjab and 20 per cent in PEPSU[13].

Secondly the provisions like self-cultivation, 'voluntary surrenders', exemption from the provisions of ceiling of land under 'co-operative garden colonies', 'well-run farms', etc., were the serious legal flaws favoring landowners and did a lot more harm to the tenants. The rate of compensation to acquire ownership of land was too high in Punjab region and was beyond the capacity of the poor tenants.

Fixation of rent at 1/3rd of the produce was favourable as compared to the traditional rate of one half. But it went against those tenants who offered to pay this rate, resulting in their eviction. The flaws in the legislation related to security of tenure of the tenants converted a benefit into loss for many tenants. In case of dispute the tenant had to fight his case in court himself which poor tenants could not afford as the judicial process is a very costly affair.

The acts imposed responsibility for implementation upon the magistrates and revenue officials-the collectors, revenue assistants, tehsildars, naib tehsildars, circle kanungo, ending with the officer closest to villagers, the *patwari*. It is a common observation that these individuals are not in a position to insure protection of the tenants. Their normal duties are complex and time consuming and they are unable to devote much time. Even if they can spare time, they are biased in favour of high landowners and many of them come from such families. The implementation machinery had been successful in effectively sabotaging land reforms.[14]

Coming to implementation of land reforms, Punjab has a very poor performance in respect of tenancy reforms. A report of the Planning Commission reads as follows:

> The law regarding security of tenure appears to be on the whole somewhat illusory. this is supported by the fact that the number of tenants in existence in the Punjab area at the time when the Punjab Security of Land Tenure Act was amended in 1955 has greatly decreased. From the information supplied by the state government on March 30, 1955 it appears that the total number of tenants including lenacts-cum-owners was 5,83,400. . . From the information supplied by the Punjab

> government recently (1964) it appears that the total number of tenants who were holding land at the commencement of the act and are still holding land as tenants is 80,250. It seems difficult to account for the large difference between these two figures. Even if we treat all the post act tenants (who were admitted up to October 1960 and are going to be settled) as those in existence in 1955 the difference between the two figures still remains large. Corresponding figures of the PEPSU area are not available but the position would not be different from that obtaining in the Punjab area[15].

Up to 1964, 19,342 tenants (in both Punjab and PEPSU) acquired proprietary rights in 1,27,794 acres of land[16]. This explains the eviction of tenants from land on a large scale. Our own study of a village in Gurdaspur district has brought out similar results[17]. Evidence collected by Wolf Ladejinsky on the study of tenurial conditions in Ludhiana supports these conslusions. According to data available, there were about 28,000 tenancies in Ludhiana district in 1953-54 and they were 40 per cent of the cultivating families (1961 Census) and cultivated 27 per cent of the total cultivated area. In 1965 the number of pure tenants was small and they were changed frequently for fear of claiming tenancy rights. In most of the cases land had been resumed by owners for self-cultivation[18]. NSS data show that area leased in as percentage of the total cultivated area in Punjab (including Haryana) declined from 35.39 per cent in 1960-61 to 28.01 per cent in 1970-71. But the decline in the number of landless tenants was much larger. They constituted 27.00 per cent of the tenants (pure tenants and owner-cum-tenants) in 1960-61 but were refused to 13.01 per cent in 1970-71.[19]

Provisions regarding the fixation of rent are not registered at all. A number of reports showed that conventional one half of the produce was paid by tenants to the landowners. Cased had been reported where rent exceeds even conventional one half of the share of produce.[20]

With the implementation of ceiling at 30 standard acres (for displaced landowners 40 acres in PEPSU and 50 acres in Punjab), about, 3,91,837 standard acres and 106 ordinary acres were declared as surplus area. Out of this 61,177 standard acres were exempted and 2,64,844 standard acres were available for disposal. Up to 1964, 1,22,862 standard acres were utilized for resettlement of 63,520 tenants but they had taken physical possession on 65,466 acres. At the time of passing of the Punjab Land Reforms Act 1972, it was officially given that 5 lakhs acres of land would be found surplus under the revised ceiling level. But up to June 1976, 14,832 returns had been scrutinized and 1957 hectares had been given. The government was expecting that a total of nearly 8,000 acres of land would be found surplus. The dismal performance of the state in this process was due to intervention of courts, procedural difficulties, pressure by the landlords' lobby and unwillingness of government to risk losing political support from the influential sections.[21] On implementation of ceiling and disposal of surplus land the best commentary is found in the findings of Harchand Singh Committee report. In July 1972, the governor of Punjab, with the approval of Punjab government, appointed a committee headed by Harchand Singh to probe into the setting up of sizeable agricultural farms on evacuee land by officers, their relatives and other influential public men. The surplus of evacuee land of various kinds after rehabilitation of refugees from Pakistan was inherited by government of India which it transferred to the government of Punjab in 1960-61 at prices considerably below their market value. It was expected that Punjab government will distribute 3,74,000 acres among the landless. This land was being cultivated by *abadkars* who had brought under cultivation the wasteland in the river beds. The government adopted the policy of open auction to dispose of this land. The influential persons through restricted auctions and illegal deals acquired a major part of this land causing eviction of poor *abadkars*. The committee after investigation found many cases of illegal deals by influential person including the agriculture minister, development minister, parliamentary secretary, the assembly speaker, Congress MPs, members and founder member of the Punjab state assembly, a former Akali minister and also a prominent leader of CPI(M), education secretary, some senior police officials and several provincial civil service officers. These influential persons had acquired land in the names of relatives, obliging tenants or landless harijans, The committee observes :

> These persons, men with means as they are, had an edge over the poor tillers, who could not mobilise requisite resources to compete with them. Even otherwise, they had the goodwill and patronage of the auctioning officers on their side. As a result, the poor and needy persons, depending upon agriculture alson, had been deprived of the opportunity of purchasing

> these lands and such actual landless tillers of land as were occupying the same, of course, were unauthorisedly uprooted. This was a tragic outcome of the policy of open auntions. It frustrated the cherished ideology of land for landless tillers[22].

The most interesting part of these findings was the fact that the very elite of the Punjab society, the individuals in commanding political positions, law makers, upholders of law or moulders of public opinion reversed the process of land reforms. These influential persons were exonerated by the state government after the resignation of ministers and speaker of legislative assembly and no follow up action had been taken.

There were two positive results of the land reforms legislation. First was the elimination of the intermediaries between the state and the cultivators. The occupancy tenants acquired ownership rights over their land. Secondly, the consolidation of holdings in the state was speedily achieved.

II

The decade of 1970s has brought remarkable changes in the land relations in Punjab. Pure tenant holdings are almost completely eliminated. They constituted 13.01 per cent of total holdings in 1970-71 but were reduced to 1.56 per cent in 1980-81. The area cultivated by pure tenants declined from 7.4 per cent in 1970-71 to 0.8 per cent in 1980-81.

Punjab agriculture continues to be dominated by owner cultivators. There is rise in the proportion of such holdings from 80.81 per cent in 1970-71 to 82.30 per cent in 1980-81 but there is decline in the area operated from 82.41 per cent to 78.77 per cent. There is 97.5 per cent increase in the owner-cum-tenant holdings during the same period. It is this category which accounts for the major part of the area under tenancy in 1980-81. Table 3.6 shows percentage distribution in the number of holdings and area under operation of different types of holding of various size groups. Data reveal that in 1970-71 the proportion of number and area of wholly owned holding increased and proportion of number and area of wholly or partially rented holdings declined with the rise in the size group of holdings.

TABLE 3.6 : Percentage Distribution of Land Tenure in Punjab in 1970-71 and 1980-81

Size-Group	*1970-71*				*1980-81*			
	No. of Holdings		*Area of Holdings*		*No. of Holdings*		*Area of Holdings*	
	Wholly Owned	*Wholly or Partly Rented*	*Wholly Owned*	*Wholly or Partly Rented*	*Wholly Owned*	*Wholly or Partly Rented*	*Wholly Owned*	*Wholly or Partly Rented*
0-1 (Marginal)	79.84	20.16	79.42	20.58	92.20	7.80	92.08	7.92
1-2 (Small)	79.84	20.16	79.81	20.19	83.87	16.13	83.02	16.98
2-4 (Medium)	80.61	19.39	80.81	19.19	79.66	20.34	79.30	20.70
4-10 (Large)	83.13	16.87	83.33	16.67	76.50	23.50	75.94	24.06
10-above (Extra large)	84.26	15.74	83.80	16.20	78.60	21.40	79.86	20.14
Total	80.81	19.19	82.41	87.59	82.03	17.97	78.77	21.23

Source : Agricultural Census, 1970-71 and 1980-81.

This was reversed in 1980-81. The proportion of the number and area of wholly owned holdings declined and proportion in number and area of wholly or partially rented holdings increased with rise in the size group. The decline in the proportion of number and area of wholly as partially rented holdings is concentrated sharply in the category of marginal and small holdings in 1980-81 over 1970-71.

In categories of medium, large and extra-large there is rise in the proportion of holdings as well as the area operated. 20.16 per cent of marginal holdings were wholly or partially rented and operated 20.58 per cent of area under the total marginal holdings in 1970-71. This declined to 7.80 per cent and 7.92 per cent respectively in 1980-81. Similarly the proportion of holdings and proportion of area of small holdings respectively declined from 20.16 per cent and 20.19 per cent in 1970-71 to 16.13 per cent and 16.98 per cent in 1980-81. There is marginal (around 1 per cent) improvement in the number and area of holdings in medium size group. But in case of large and extra large groups the increase is quite high. In case of large and extra large groups the increase is quite high. In case of large holdings the proportion of number and area increased respectively from 16.87 per cent and 16.67 per cent in 1970-71 to 23.50 per cent and 24.06 per cent in 1980-81, and in case of extra large group from 15.74 per cent and 16.20 per cent respectively in 1970-71 to 21.40 per cent and 20.14 per cent in

1980-81. This indicates that the proportion of holdings is rising. This shows that tenancy is shifting from small operators to large operators. In fact land is shifting from the small holders to large holders-a capitalist development tendency of concentration of productive resources in hands of a few producers of large size. This is revealed by Table 3.7. There is a decline of pure tenant holdings by 91.0 per cent during 1970-71 to 1980-81. The decline is almost even over different size class. In case of owner-cum-tenant holdings there is a rise in the number by 97.5 per cent which is fairly;y uneven among different size groups.

The largest rise is experienced by medium holdings followed by large, small and extra large and marginal holdings. Marginal holdings rose by 17.6 per cent whereas medium holdings by 116.8 per cent and large holdings by 112.1 per cent, the small holdings rose by 90.9 per cent, extra large by 83.9 per cent. This is further confirmed by the change in the number of wholly ownership holdings which experienced a net decline by 24.7 per cent during 1970-71 to 1980-81. This is mainly experienced by marginal and small holdings which respectively declined by 55.8 per cent and 19.5 per cent. The medium and extra large holdings increased respectively by 1 per cent and 0.2 per cent and large holdings declined by 2,9 per cent. This indicates that relative position of extra large, large and medium farmers in terms of number has improved over marginal and small holdings. This is further confirmed by changes in the land operated by different class size of farmers. Table 3.8 shows that there is decline in the land operated by wholly owned holdings to the extent of 6.3 per cent during this period whereas extra large holdings is the only category which gained land by 1.5 per cent, the marginal holdings experienced decline by 39,1 per cent, small holdings by 12,6 per cent, medium holdings by 2.4 per cent and large holdings 5.8 per cent. Owner-cum-tenant holdings experienced a considerable gain in the land operated by them. The gain in area to all such holdings was to the extent of 95.8 per cent. The gain is largest to large holdings which experienced a rise in the land operated by 111.7 per cent followed by medium holdings with 110.5 per cent, small holdings by 92.6 per cent and extra large holdings like their number, experienced a decline in the area operated by 89.0 per cent. The loss of area by different size class follows the same pattern as that of the loss in the number of holdings. Though there is not much difference among different size groups the loss is more with marginal holdings and extra large holdings compared to small, medium and

TABLE 3.7 : Percentage Variation in the Number of Land Tenure in Punjab in 1980-81 over 1970-71

Size Group	*Number of Wholly Owred*		*Percentage Variation*	*Number of Owner-cum-Tenant*		*Percentage Variation*	*Number of Pure-Tenant Holding*		*Percentage Variation*
	1970-71	*1980-81*		*1970-71*	*1980-81*		1970-71	1980-81	
Marginal	413204	182608	–55.8	8557	10064	17.6	95807	5388	–94.4
Small	207616	167211	-19.5	14488	27668	90.9	37949	4487	–88.2
Medium	226585	228965	1.0	24907	54013	116.8	29611	4445	–84.9
Large	205948	199835	–2.9	28179	59783	112.1	13628	1583	–88.4
Extra large	58038	58124	0.2	8500	15638	83.9	2345	178	–92.4
Total	1111421	836743	–24.7	84631	167166	97.5	179340	16081	–91.0

Source : *Agricultural Census*, 1970-71 and 1980-81.

TABLE 3.8 : Percentage Variation in the Area of Land Tenure in Punjab in 1980-81 over 1970-71

Size Group	*Wholly Owned Holdings*		*Percentage Varation*	*Owner-cum-Tenant*		*Percentage Variation*	*Pure Tenant Holding*		*Percentage Variation*
	1970-71	1980-81		1970-71	*1980-81*		*1970-71*	1980-81	
Marginal	178893	108863	–39.1	5302	6204	17.0	41059	3163	–92.3
Small	266976	233204	–12.6	21404	41219	92.6	53749	6477	–87.9
Medium	642576	627171	–2.4	72216	151986	110.5	80415	11698	–85.4
Large	1264858	1189041	–5.8	173848	368043	111.7	78517	8592	–89.0
Extra large	894489	9077796	1.5	132747	226449	70.0	40136	8286	–94.3
Total	3264612	3066075	–6.3	405519	791174	95.8	293878	32216	–88.0

Source : *Agricultural Census, 1970-71 and 1980-81.*

large holdings. As a consequence of these changes, the total number of pure tenants and owner-cum- tenant holdings declined from 2,63,971 in 1970-71 to 1,83,247 in 1980-81 but area operated by them increased from 6,99,397 hectares to 8,26,387 hectares. The average size of such holdings was 2.46 hectares which was lower than 2.89 hectares the overall average size of landholdings in the state in 1970-71 increased to 4.51 hectares in 1980-81 which became higher than 3.79 hectares, the overall average size of the land holding in the state.

The available evidence suggests a growing tendency towards capitalist agriculture in the state. There is overall decline in the total numberof holdings in the state by 25.3 per cent during 1970-71 and 1980-81. There is almost negligible decline in the number of individual holdings but there is a high decline in the number of joint holdings from 3,78,933 to 36,222 a decline by 90.5 per cent. This decline with some variation is experienced by all size classes of holdings. The overall decline in the total number of holdings of all types is not evently experienced. Whereas the marginal holdings declined by 61.9 per cent and small holdings by 23.3 per cent, the medium holdings gained by 2.3 per cent,large holdings by 8.6 per cent and extra large by 7.3 per cent. The same pattern in change is observed in case of individual holdings as well as the area operated. There is little evidence to suggest the large scale of land by small and marginal farmers.[23]

Evidence available from various studies indicates that all class size of farmers participate in land market over purchase and sale of land is the state. Small farmers buy and sell land as other classes of farmers do. In the process land transfer takes place but there is a marginal inter-class transfer of land. A survey of Ghanour block of Patiala district reveals the land transfer between 1973 and 1978 as given in Table 3.9.

Table 3.9 indicates that out of the total 482 acres of land sold during the five-year period only 12 acres have experienced inter-class transfers. This transfer is from both the categories, i.e., small and medium farmers to large farmers and accounted for 2.49 per cent of the total land sold and purchased. The remaining 97.51 per cent of the land sold/purchased have remained within the respective classes. This is confirmed by our own study of three villages each from Patiala and Gurdaspur districts.[24]

TABLE 3.9: Sale Purchase of land in Ghanour 1973-78

(in acres)

Ownership Holdings	*Sale*	*Purchase*	*Net Increase*
Small (up to 5)	62 (12.86	53 (11.00)	–9
Medium (5 to 15)	174 (36.10)	171 (35.47)	–3
Large (above 15)	246 (51.04)	258 (53.35)	+12
Total	482 (100)	482 (100)	

Notes : Figures in parentheses are in percentage.

Sources : Baldev Singh, *Land Market: Theory and Practice in India,* Agricole, New Delhi, 1982, pp. 67-68.

The evidence suggests that the shift of land from the small and marginal farmers to medium and large farmers is on account of leasing out of land. A recent study on tenancy relations has revealed that in the three sampled villages of Ludhiana district and two villages in Faridkot district in 1979-80 a sizeable percentage of tenants are big cultivators with a substantial land area of their own, fully endowed with modern production assets such as tractors, tubewells, etc, enterpreneurial skills, high commercial status.[25] This is further confirmed by another study of a Ludhiana village. It reveals that most of the leasing-out is done by small owners and leasing-in by the big operators.[26]

Implications for Land Reforms

The changes in land relations are the outcome of changes in mode of production of Punjab agriculture. They correspond to capitalist mode of production. The new pattern of tenancy is capitalist in nature. This pattern of land relations has serious implications for land reforms devised to change precapitalist relations of production. The land reforms in India as devised in 1950s and extended in the later period were supposed to change obsolete land relations which were obstructing expansion of forces of production on capitalist lines. Once land relations are in tune with the new mode of production further implementation of these reforms lose their social significance. This is evident from the emerging situation of Punjab. The implimentation of Tenancy Act providing security of tenure will lead to transfer of land from the small owners to large owner who are leasing in land from the former to have high size holding for cultivation on capitalist lines.

Originally this act was meant for protection of small tenants cultivating land of big landlords. Now its operation will do the opposite.

Secondly, implementation of ceiling laws and acquisition of surplus land from the big owners, its distribution among the landless to make them small cultivators will lose purpose. This is because within dynamics of Punjab agriculture, land is shifting towards large cultivators through new pattern of tenancy. Whereas operation of tenancy laws can go against the poor and small owners of land, the operation of ceiling act as such and surplus distribution of land to the landless no longer remains capable of reducing inequality in land distribution. The situation requires necessary modifications in these acts to protest the weaker farmers.

First, operation of Tenancy Act for protection of tenants against eviction must be restricted for small cultivators. Its use must not be allowed for large cultivators. Secondly for usefulness of ceiling laws and land redistribution among the poor, there is a need to simultaneously take steps to check tendency of small cultivators going out of operation. Here there can be two lines of thinking: (a) small producers should be protected through various measures of the government by providing assistance and facilities serving exclusively the needs of such farmers. The experience of the past shows that is spite of such measure during the decade of 1970s large proportion of marginal and small farmers have gone out of cultivation. A large number of programs meant for these farmers are in fact helping the medium and large farmers. In absence of strong organisations of such farmers these programmes cannot protect them. The situation immediately requires some sort of co-operative effort for protection of poor/small cultivators. The fact that there is a large decline in the joint holdings indicates that with the growth competition has been strengthened and community co- operative spirit has weakened. In this situation it is difficult to visualise the success of co-operative farming by the small cultivators. The widespread corruption in co-operative departments (like other areas of public life) further adds to the difficulties. The only way to check liquidation of poor/small cultivators is to introduce land reforms on Soviet and Chinese pattern.[27]

In these days there is a lot of talk at the highest level about the strengthening of institution of Panchayati Raj. Institution of Gram Sabha/Gram Panchayat can be assigned the historic role of re-organisation of rural life.[28] It can be made the instrument of collectivisation by enthrusting ownership of entire land of the village. It can be made responsible for further modernisation of agriculture and organisation of non-agricultural activities in the village. In the present context modern rural industries particularly those engaged in processing of agricultural products can be started at village level. this pattern of organisation of rural society has a high potential for modernisation and is capable of resolving the major problems created by the development process. It is some form of collectivisation which can protect the poor cultivators from the broad dynamism of capitalist operations in agriculture. It is in this sense that land reforms are relevant and more urgently required in the state.

Note : Reproduced from, *Economics and Political Weekly*, June 24, 1989 with kind permission of the publisher.

NOTES

(Comments on an earlier draft by Surjit Hans and A. S. Dhesi are greatefully acknowledged).

1. B. D. Talib, *Agrarian Tension and Peasant Movement in Punjab* (mimeo), Centre for Study of Regional Development, Jawaharlal Nehru University, New Delhi, 1979.
2. Wolf Ladejinsky, 'Field Observation in the Punjab' in *Selected Paper of Wolf Ladejinsky*, edited by L. J. Walinsky, Oxford University Press, New Delhi, 1977.
3. Pranab Bardhan, 'Variations in Extent and Forms of Agricultural Tenancy-I: An Analysis of Indian Data across Regions and Over Time, *Economic and Political Weekly* September 18, 1976.
4. Pranab Bardhan, 'Variations in Extent and Forms of Agricultural Tenancy-II' *Economic and Political Weekly*, September 18, 1976.
5. See Table 3.1.
6. Wolf Ladejinsky, *op cit.*
7. Government of PEPSU, 'Land Reforms in PEPSU', *Indian Journal of Agricultural Economics*, Vol. VII, No. 1, March 1953.
8. R.L. Bhalla, *An Economic Survey of Bairampur,* Government Press, Lahore, 1923, quoted in M.L. Darling, *Punjab Peasantry in Prosperity and Debt.*
9. The Bord of Economics Inquiry, Punjab, *Reservey of Village Suner, District Ferozepur*, Government of Punjab, 19963, p 41.
10. The Board of Economic Inquiry, Punjab, *Reservey of Village Gijhi, District Rohtak*, Government of Punjab, 1960, p. 56.
11. *Punjab Land Reforms Act, 1972* Sections 4, 14 and 27.
12. Wolf Ladejinsky, *op. cit.*
13. Report of the Committee on Tenancy Reforms, Planning Commission, 1956.
14. Wolf Ladejinsky, *op cit.*

15. D. Bandhopadhyaya, 'Indra Lohar and the Due Process of Law', *Economic and Political Weekly,* September 30, 1972, Planning Commision, Report of the Task Force on Agrarian Relations, Government of India, 1973.
16. Pranab Bardhan, *op. cit.*
17. Sucha Singh Gill, 'Agrarian Transformation in a Punjab Village', *Mainstream*, March 4, 1978.
18. Wolf Ladejinsky, *A Study of Tenurial Conditions in Package Districts*, Planning Commission of India, 1966.
19. Pranab Bardhan, *op. cit.*
20. Planning Commission (1966), *op. cit*; Wolf Ladejinsky (1965), *op. cit*; Wolf Ladejinsky, Green Revolution in Punjab—A Field Trip, *Economic and Political Weekly*, June 28, 1969 Sucha Singh Gill, *op. cit.*
21. 'Little Headway in Land Reforms', *Economic and Political Weekly*, July 10, 1976.
22. *Harchand Singh Committee Report*, Punjab Government, 1973, p. 71.
23. H.S. Shergill, 'Land Sales and Land Prices in Punjab: 1952-53 to 1978-79, *Economic and Political Weekly*, Review of Agriculture, September 20-27, 1986. His arguments of stabilisation effect of green revolution on small and marginal farmers is valid. Calculations from his Table 3.1 about average annual area sold reveal that in pre-green revolution period (1952-53 to 1965-66) it was 150 acres and in post-green revolution period 1966-67 to 1978-79 it increased to 206.9 acres, leading to increased land sales due to modernisation in the sample villages of Sangrur district of Punjab.
24. Sucha Singh Gill, 'Agrarian Structure of Punjab' in Amit K. Gupta, *Agrarian Structure and Peasant Revolt in India*, Criterion Publishers, New Delhi, 1986.
25. Iqbal Singh, *Changing Tenancy Structure in a Developing Agricultural Region—A Case Study of Punjab,* unpublished Ph.D. Dissertation, CRD, JNU, New Delhi, 1985, Chapter 6. Quoted in G. K. Chadha, *The State and Rural Economic Transformation: The Case of Punjab* 1950-85, Sage Publications, New Delhi, 1986, p. 227.
26. S. S. Grewal and P. S. Rangi, 'An Analysis of Agrarian Structure in Punjab, *Indian Journal of Agricultural Economics, October-December 1981.*
27. Erich H. Jacoby, 'Has Land Reform Become Obsolete', H.J. Hobsbawm *et al* (eds), *Peasants in History*, Oxford University Press, New Delhi, 1980.
28. Our attention to this issue has been drawn by Suresh Ramabhai.

4

LAND MARKET TRANSACTIONS AND EXPANSION/CONTRACTION OF OWNED AREA OF CULTIVATING PEASANT FAMILIES IN THE PUNJAB

H. S. SHERGILL

In a commercialised agrarian community economic competition among peasant families occurs mainly in the land market. Persistent superiority/inferiority in economic performance of a peasant family ultimately gets reflected in a change in its owned land and peasant household move on the economic ladder mainly via purchase/sale of land. This continuous redistribution of farm land amongst peasant families via the land market is supposed to be the main mechanism through which the amorphous peasant mass gradually gets differentiated into classes over time. This paper presents the results of an empirical exercise carried out to isolate some key factors that influence the expansion and contraction of owned land of peasant families via land market transaction. The impact of these land market transaction on the distribution of farm land amongst cultivating peasant household is also analysed. For convenience of usage peasant farms are called here *growing* and *decaying* respectively if their owned land base has expanded and contracted as a result of purchase and sale of land since 'birth'. Those that did not transact any land in the market since 'birth' are labelled as *static farms.* The entry point of a peasant family farm in economic competition is reckoned from the year of its 'birth' as an independent socio-economic unit and 'birth' is defined from the time point when the present head of the household took charge of family farm affairs.

GROWTH AND DECAY OF FARMS : SOME CAUSAL FACTORS

The processes of growth and decay of peasant farms are very complex because of the fusion of 'family' and 'farm' in a single composite socio-economic entity and it is not easy to discern why some peasant families are able to accumulate wealth and buy land, while many others fall in chronic debts and sell land. In this context, Marxist scholars give primacy to the initial size of a farm in determining its fate in economic competition.[1] It is argued that peasant families that enter economic competition with a sizeable land area of their own are able to take advantage of advanced technology, better terms in credit and marketing arrangement and many other economies of scale and are thus able to generate bigger production surpluses, accumulate wealth and buy more land. Their moneylending/trade operations and political clout are also supposed to help these big owners in the accumulation of wealth and acquisition of land. On the other hand, use of inferior technology, disadvantages in credit and marketing arrangements and the sheer inefficiency and vulnerability of a small sized family enterprise, combined with the excessive pressure of consumption needs are supposed to ruin most of the small owners, resulting ultimately in the sale of their lands.

Schumpeter has, however, cautioned against assuming any such 'automatism of accumulation' in big family enterprises and has suggested that there may rather be a tendency towards 'automatic decline' (see Schumpeter, 1951, pp. 148-176). In his view 'initial big size' in itself does not guarantee the expansion of a family enterprise, notwithstanding the usual advantages accruing from the already elevated economic position of such a unit. Bigger production surpluses do not get translated into successful investment automatically, but must be deliberately and wisely invested, and must not be squandered in consumption. Schumpeter has rather emphasised the primary role of subjective factors in the success/failure of family enterprises in economic competition. It is the differences in disposition to save, quality of management and leadership of enterprise, capacity to do prolonged hard work and renounce other pleasures of life, ability to adopt new techniques quickly and the sheer hard-headedness and drive to expand the enterprise at any cost, that distinguish the successful families from the not so successful. He even discounts the role of 'chance' in success/failure of family enterprises by the argument that what matters is not 'chance' in itself, but rather how a family adapts to or exploits its consequences.

Chayanov and other neo-populists have emphasised the primary role of demographic factors like family size in the expansion and contraction of present farms over time (see Chayanov, 1966 and Shanin, 1972). According to this view, farm size adapts to family size in the long run; bigger families gaining land and smaller ones losing it through periodic communal repartitioning of land in pre-capitalist agrarian communities. The mechanism through which family size influences the size of owned land of a family in a land regime in which private property rights in land are firmly established, however, was never explicitly stated by neo-populists. Causal influences from family size to farm size can be many and may even work at cross-purposes. One that can be thought is that big family size and large number of adult males in it may not only create the motivation for acquiring more land to provide 'inheritance', work and 'sustenance' to members, but also help in its achievement by enhancing a family's accumulation ability. Bigger families enjoy considerable advantage in saving due to reasons of family public goods, scale economies and complementality in consumption and also from specialisation and division of labour in household duties.[2] A negative effect of big family size on accumulation ability is, however, also plausible; consumption pressure generated by bigger families is supposed to ruin many a small owners resulting ultimately in the sale of their lands.

In addition to these factors suggested by well-known theories, some others like education of family labour force, financial burden resulting from addiction of family males to wine/opium and other wasteful habits, and 'financial balance' inherited from the parent unit at the time of 'birth' are also likely to significantly influence the chances of survivability and growth of peasant farms. Education may enhance the chances of growth in many ways like early adoption of new technology and of rational production and accounting practices, and also by rousing the ambition of family males for expansion of their farm. Addiction to wine/opium of family males increases the financial burden on a family tremendously and also weakens the work capacity and 'will' to retain land of such addicted members, thus adversely affecting the family economy from many sides.[3] 'Financial balance' inherited from the parent unit is also likely to play an important role in the economic fate of a family in its subsequent career; those starting with debt inherited from the parent units being more prone to decay via land sales, and those that start with an inheritance of cash savings from the parent units being in a better position to accumulate more and buy land.

The role of these above-mentioned factors may be very important, but what ultimately matters in the growth and decay of peasant farms is the ability of a peasant family to meet its consumption requirements from the current flow of income from farm and non-farm sources. Persistent failure to do so results in chronic indebtedness and ultimately to forced sale of land. Success in meeting the consumption requirements of family members provides financial stability; and ability to generate incomes (from farm and non-farm sources) even larger than the consumption requirements helps in the accumulation of savings and wealth that may ultimately be used to purchase more land. This suggests that per capita income of the family is likely to be one of the most important determinants of expansion/contraction of its owned area through land market transactions.

DATA AND METHODOLOGY

The data used in this study were collected by the author during 1979-80 from 90 peasant farms selected from 14 villages of Sangrur district of the Punjab, through stratified random sampling procedure. Samples of 30 growing farms and 30 decaying farms were drawn from the list of buyers and sellers of land recorded in the Mutation Registers of the 14 villages for the year 1978-79, and the sample of 30 static farms was drawn from the list of all cultivating households maintained in the House-tax Register of each village. From these 90 peasant households details of their family histories, farm economies and sale/purchase of land since 'birth' were collected during 1979-80.

Methodologically, the question of growth and decay of peasant farms can be viewed from many different angles. One approach can be to consider not only the growth and decay processes separately from each other, but also the triggering of these processes separately from the subsequent quantum of land gained or lost through land market transactions. Another, more direct and simpler approach can be to view growth and decay as two poles of the same single process of change in the owned area of peasant families through land market transactions. In this paper we have adopted this direct approach and have tried to capture the growth and decay of farms by defining a single variable, namely, absolute change in the owned area (in acres) through land market transactions since the birth of a household as an independent unit. This variable is then used as the dependent variable in the regressions.

Specification of Explanatory Variables

Keeping in view the availability of information in our data set and the importance of different factors in the growth and decay of peasant farms, the following explanatory variable were defined:

1. *Area owned at birth :* The area of land owned (in acres) by the household at the time of its 'birth' as an independent unit. It represents the 'initial farm size' factor emphasised by Marxist scholars and is expected to have a positive influence on change in the owned area via land market transactions.

2. *Number of adult males :* The number of adult male members in the family. It represents the neo-populist family size factor and is expected to influence the change in the owned area via the land market positively.

3. *Education :* The percentage of adult male members in the family with eight or more years of schooling. It is expected to take a positive sign in the regression model.

4. *New technology adoption :* The number of years elapsed since the adoption of Maxican high-yielding variety (HYV) wheats up to the year of the survey, i.e., 1979-80. It takes higher values for early adopters and lower values for late adopters and is expected to influence the change in the owned area positively.

5. *Financial position at birth :* The amount of mortgaged in/mortgaged out land inherited from the parent unit at the time of 'birth' of the household. It is in acres and takes positive sign if the household inherited 'mortgaged in' land, negative sign if it inherited 'mortgaged out' land and zero value if did not inherit either of the two. It is expected to have a positive influence on change in the owned area via land market transactions.

6 *Per capita income :* It is defined in rupees per year and is computed by dividing the total family income (from farm and non-farm sources), with the number of standardised consumers in the family. [4] It is expected to have a positive sign in the regression model.

7. *Moneylending :* The actual amount of lending per family could not be used for want of reliable information on that; and we settled for the second best and defined this variable as a dummy taking values

of '1' and '0' respectively for households practising and not practising money lending.

8. *Addition :* In defining this variable the average expenditure incurred on wine/opium, etc., could not be used on account of absence of reliable information on that. So this variable was also defined as a dummy taking values of '1' and '0' respectively for families having and not having same male members chronically addicted to wine/opium.

EMPIRICAL RESULTS

The results of tests on the differences in means/proportions of explanatory variables among growing, decaying and static farms are presented in Table 4.1 and univariate and multivariate regression results on determinants of change in the owned area through land market transactions are given in Table 4.2. Since money lending and addiction variables have been defined as dummies for want of reliable information on the amount lent and the average expenditure on wine/opium, regressions were estimated alternatively by including as well as excluding these variables. Significant influence of area owned at birth, the number of adult males, financial position at birth, per capita income, money-lending and addiction on change in the owned area through land market transactions is clearly revealed by the results presented in Tables 4.1 and 4.2. The influence of new technology adoption on change in the owned area through land market transactions is indicated by the difference in means test and univariate regression coefficient of this variable, but is not very clearly confirmed in the multivariate context; its coefficient has the expected positive sign in both the multivariate regressions, but is significant at 10 per cent level in only one of the equations. Taking an overall view of empirical results given in both the tables and the unquestioned importance of new technology adoption in production efficiency and farm incomes, it may not be out of order to conclude that new technology adoption is also an important determinant of change in the owned area through land market transactions. The difference in means test result (Table 4.1) mildly alluded to the positive influence of education on change in the owned area via the land market, but it failed to come out in the univariate/multivariate regression results given in Table 4.2; even the sign of its coefficient became negative in one of the regression equations. One possibility is that education has no direct influence on change in the owned area through the land market, but its influence

TABLE 4.1 : Growing Decaying and Static Farms : Univariate Comparison of Determinants

	Variable	*Mean/Proportion for*			*Mean/Proportion differences*			*F-value/ Chi-squared*
		Growing (G)	*Decaying (D)*	*Static (S)*	*(G-S)*	*(D-S)*	*(G-D)*	
	1	2	3	4	5	6	7	8
1	Area owned at birth (acres)	13.67	14.86	15.05	–1.38 (0.51)	0.19 (0.07)	–1.19 (0.44)	0.15
2	Number of adult male members	3.43	2.47	3.03	0.40 (1.04)	–0.56 (1.47)	0.96[b] (2.51)	3.18[b]
3	Proportion with some male members having ≥ 8 years schooling	57%	27%	40%	0.17 (1.34)	–0.13 (1.10)	0.30[b] (2.25)	5.60[b]
4	Years elapsed since the adoption of Mexican wheats (counted from the year of survey, 1979-80)	12.63	10.80	11.20	1.43[a] (2.91)	–0.40 (0.81)	1.83[a] (3.72)	7.66[a]
5a	Proportion inheriting wealth in the form of 'mortgaged in' land from the parent unit	17%	0%	3%	0.14[c] (1.98)	–0.03 (0.42)	0.17[b] (2.40)	9.59[c]
5b	Proportion inheriting debt in the form of 'mortgaged out' land from the parent unit	3%	43%	7%	–0.04 (0.40)	0.36[a] (3.56)	–0.40[c] (3.96)	20.20[a]
5c	Acres of mortgaged in / mortgaged out land inherited from the parent unit	0.33	–1.60	–0.17	0.50 (1.13)	–1.43[a] (3.22)	1.93[a] (4.35)	10.19[a]
6	Per capita income from all sources (Rs.)	4,121	1,358	2,799	1322[a] (2.70)	–1441[a] (2.95)	2763[a] (5.65)	15.97[a]
7	Proportion practising moneyending	23%	0%	0%	0.23[b] (2.58)	0.00	0.23[b] (2.58)	15.15[b]
8	Proportion with some male members addicted to wine/opium	3%	33%	3%	0.00	0.30[a] (3.00)	–0.30[a] (3.00)	15.58[b]
9	Age of household (number of years elapsed since birth as an independent unit up to the year of survey, i.e., 1979-80	18.43	16.37	14.83	3.60 (1.30)	1.54 (0.59)	2.06 (0.63)	0.85

Notes : 1. Figures in parentheses are t-values for means and z-values for proportions.
2. a, b, c, indicate respectively significant at 1 per cent, 5 per cent and 10 per cent levels for a two-tailed test.

TABLE 4.2 : Change in Owned area through land market : Regression results on determinants

(N=90. Dependent variable : Change (in acres) in area owned at birth through land market transactions)

Explanatory variable	*Univariate regression coefficient*	*Multivariate regression results*			
		A		B	
		Regression coefficient	*Beta coefficient*	*Regression coefficient*	*Beta coefficient*
1	2	3	4	5	6
1. Area owned at birth	−0.013	−0.072[b]	−0.171	−0.108[a]	−0.257
	(0.29)	(2.09)		(2.99)	
2. Number of adult males	0.724[b]	0.474[b]	0.167	0.611[a]	0.215
	(2.48)	(2.39)		(2.78)	
3. Education	0.020	0.001	0.009	−0.009	−0.056
	(1.20)	(0.11)		(0.63)	
4. New technology adoption	0.526[b]	0.187	0.087	0.285[c]	0.132
	(2.38)	(1.14)		(1.67)	
5. Financial position at birth	0.761[a]	0.429[b]	0.173	0.552[a]	0.223
	(3.03)	(2.42)		(2.83)	
6. Per capita income	0.001[a]	0.009[a]	0.423	0.001[a]	0.618
	(6.56)	(4.53)		(7.33)	
7. Money-lending	9.910[a]	5.818[a]	0.358		
	(7.21)	(4.53)			
8. Addiction	−3.751[a]	−1.554[c]	−0.121		
	(2.87)	(1.67)			
Intercept		−4.137		−5.699	
R-SQR		0.620		0.511	
R-BAR-SQR		0.582		0.476	
F-value		16.51		14.47	

Notes : A includes all explanatory variables. B excludes moneylending and addiction variables.

1. Figures in parentheses are t-values.
2. a, b and c indicate respectively significant at 1 per cent, 5 per cent and 10 per cent levels for a two-tailed test.

operates mainly via inducing early adoption of new technology and generating higher per capita income from farm and non-farm sources (see correlation matrix given in Table 4.3). Another possibility is that the influence of education on change in the owned area via the land market operates more complexly and our linear specification fails to reveal that.[5]

Except for area owned at birth variable, the direction of influence of all the other variables found statistically significant in the regressions, is in line with their hypothesised effects. The observed negative coefficient

TABLE 4.3 : Correlation Matrix

Variable	X_1	X_2	X_3	X_4	X_5	X_6	X_7	X_8	X_9	X_{10}
1	2	3	4	5	6	7	8	9	10	11
1. Area owned at birth (X_1)	1.000	0.066	0.319[a]	0.232[b]	–0.082	0.351[a]	0.022	0.304[a]	–0.031	0.661[a]
2. Number of adult males (X_2)		1.000	0.040	0.118	0.030	0.059	0.135	–0.085	0.255[b]	–0.237[b]
3. Education (X_3)			1.000	0.381[a]	0.141	0.282[b]	0.002	0.099	0.127	0.154
4. New technology adoption (X_4)				1.000	0.114	0.231[b]	0.147	–0.041	0.246[b]	–0.023
5. Financial position at birth (X_5)					1.000	0.080	0.129	–0.198[c]	0.308[a]	–0.191[c]
6. Per capita income (X_6)						1.000	0.434[a]	–0.065	0.573[a]	0.109
7. Moneylending (X_7)							1.000	–0.114	0.609[a]	–0.092
8. Addiction (X_8)								1.000	–0.293[a]	0.242[b]
9. Change in owned area via market (X_9)									1.000	–0.224[b]
10 Per male member area owned at birth (X_{10})										1.000

Note : a, b and c indicate significant at 1 per cent, 5 per cent and 10 per cent levels respectively for a two-tailed test.

of area owned at birth is contrary to our expectation and directly conflicts with the widely held view (particularly among Marxist scholars) that the peasant families which start with a larger owned area fare better in economic competition and those that enter the economic race with a smaller owned area are more likely to fail and get ruined financially. In view of this contradiction, it is necessary to check the validity of this negative coefficient with data sets originating in other time-space settings in the Punjab. In the specific case of our sample a number of factors seem to be responsible for this outcome. One is the significant high positive correlation between area owned at birth and per male member area owned at birth (see Table 4.3), which indicates that the pressure to acquire more land was higher on families with smaller area owned at birth and lower on those starting with larger owned area at birth. In a peasant cultural environment, the pressure to acquire more and retain the already owned land is likely to be inversely related with per male member area owned. Significant positive correlation of area owned at birth with addiction variable (see Table 4.3) also indicates that families starting with larger owned area at birth are more prone to acquire expensive consumption habits and even have some male members chronically addicted to wine/opium. It may also be noted from the correlation matrix given in Table 4.3 and multiple regression results reported in note 6 that addiction variable is not significantly correlated with per capita income and even the sign of their relation is negative.[6] Given these results, it is difficult to accept the usually made statement that higher incomes resulting from Green Revolution are causing addiction in the rural Punjab. At least in our sample that does not seem to hold. The legal constraints created by the existing ceiling laws may have also contributed in generating this tendency of inverse relation between the size of area owned at birth and change in it through land market transactions. Households already owning considerable land have little incentive to purchase more and may have even disposed off some land through the market for fear of losing it through the operation of ceiling laws.[7] All these explanations of the observed inverse relation of area owned at birth with change in it through land market transactions are at best only preliminary hypotheses and need more detailed examination. Even the existence of this inverse relation needs to be confirmed with data sets originating in other time-space settings, before accepted as a valid generalisation.

Impact on Land Distribution

We next investigated the impact of these land market transactions of the sample of growing and decaying farms on the distribution of land among rural land owning households. Only the change in the area owned by growing and decaying farm households was compared for this purpose, as the area owned by static farm households has not changed at all due to their non-involvement in land market transactions. The relevant information is summarised in Table 4.4. It reveals a net transfer of land from the non-cultivating owners to the cultivating owners via land market transactions; the excess of 67.6 acres in the total land purchased by growing farms group over that sold by decaying farms group indicates the approximate quantum of this net transfer of land from the non-cultivating owners to the cultivating owners.[8] Further, a perusal of columns (6) and (7) of Table 4.4 suggested that though all the three size groups of cultivating owners (small, medium and large) made net gains of land via market transactions, the biggest gain was made by the small owner cultivators' group; the area owned at birth by households of this group increased by as much as 20.18 per cent as a result of their land market transactions. A conclusion that may be safely drawn from these data is that land market transactions of the sample farms have not resulted in greater inequality in the distribution of land among cultivating owner households; rather a mild centripetal tendency is revealed by the small increase in the share of owned land of the small owner cultivators' group, and a slight decrease in the share of large owner cultivators' group. It may be recalled that the significant negative coefficient of area owned at birth variable in the regression results discussed earlier, also alluded to a centripetal tendency of this type.

CONCLUSIONS

The main conclusions of our analysis of land market transactions of a sample of farms from Sangrur district of the Punjab may now be recapitulated :

1. The change in the owned area of peasant families through land market transactions is positively influenced by per capita income, number of adult males in the family, financial position at birth,

TABLE 4.4 : Market Transfers of Land and Change in the Distribution of Land Among Cultivating Households

Size-groups (acres) by land owned at birth	*Number of growing and decaying farms*	*Area owned at birth*	*Area purchased since birth*	*Area sold since birth*	*Net gain/loss of land via land market*		*Area owned at present*
					In acres (col. 4–col. 5)	*As per cent of area owned at birth* $\left(\frac{col.\ 6}{col.\ 3} \times 100\right)$	
1	2	3	4	5	6	7	8
Small	27	148.2	60.1	30.2	29.9	20.18	178.1
(0-9.99)	(45)	(17.33)	(38.48)	(34.09)	(44.45)		(19.30)
Medium	18	272.5	47.6	27.2	20.4	7.5	292.9
(10.00–19.99)	(30)	(31.87)	(30.47)	(30.70)	(30.18)		(31.75)
Large	15	434.3	48.5	31.2	17.3	4.0	451.6
(20.00 and above)	(25)	(50.80)	(31.05)	(35.21)	25.59		(48.95)
All/Total	60	855.00	156.2	88.6	67.6	7.91	922.6

Note : Figures in parentheses indicate percentage to the total.

moneylending and early adoption of new technology; and is negatively influenced by addiction of family males to wine/opium and large size of owned area at birth.

2. Land market transactions seem to be resulting in a net transfer of land from non-cultivating owners to cultivating owners and also produce a mild centripetal tendency in the distribution of land amongst this latter group of owner cultivators.

Note : Reproduced from, *Indian Journal of Agricultural Economics*, Vol. 45, No. 1, Jan-Mar. 1990 with kind permission of the publisher.

NOTES

1. For a classic statement of this standard Marxist thesis, see Lenin (1971).
2. For examples and discussion, see Lazear and Michael (1980).
3. For one of the earliest recorded evidence on the rold of addiction in land sales in the Punjab, see Punjab Board of Economic Enquiry (1925).
4. Net farm income was defined as follows:
   Gross value of crops and milk produced—(Value of all material inputs used + cost of purchased primary factors + imputed interest on credit financed for the purchase of modern inputs + imputed depreciation on fixed capital). The conversion formula used for estimating the number of standardised consumers was taken from the Government of Punjab (1978).
5. Some further probing that we did to analyse the role of education in the growth and decay of peasant farms confirmed this complexity and suggested that education significantly increases the chances of a family buying land, but does not influence significantly its chances of selling land.
6. To further check the influence of per capita income and area owned at birth variable on addiction, the following regression equation was estimated :

$$\begin{aligned} \text{addiction} = {} & 0.234 - \underset{(1.61)}{0.00003} \text{ per capita income} \\ & + \underset{(3.11)}{0.0114^{a}} \text{ area owned at birth} - \underset{(0.86)}{0.01916} \\ & \text{number of adult males} + \underset{(0.85)}{0.00121} \text{ education} \\ & - \underset{(0.81)}{0.01496} \text{ new technology adoption} \end{aligned}$$

7. More detailed analysis of the role of ceiling laws could not be done, because we had not collected information on the reasons given by the peasant families themselves for not purchasing/selling land.
8. It may be noted that our sample was restricted strictly to cultivating owner households. The difference in the area purchased by growing farms and that

sold by the decaying farms is due to the area gained by the cultivating-owner group from the excluded group of non-cultivating owners.

REFERENCES

Chaynov, A.V. (1996). The Theory of Peasant Economy, Richard D. Irwin, Homewood, Illinois, U.S.A.

Government of Punjab (1978). Report of the Working Class Family Income and Expenditure Survey in Punjab, 1975-76, Economic and Statistical Organisation, Punjab, Chandigarh.

Lazear, E.P. and R.T. Michael (1980) . "Family Size and the Distribution of Real Per Capita Income". *The American Economic Review,* Vol. 70, No. 1 March.

Lenin, V.I. (1971) *The Development of Capitalism in Russia,* Progress Publishers, Moscow.

Punjab Board of Economic Enquiry (1925). An Enquiry into Mortgages of Agricultural Land in Kot Kapura Uttar Assessment Circle of Ferozpur District, Lahore.

Schumpeter, J.A. (1951). *Imperialism and Social Classes,* Basil Blackwell, Oxford

Shanin, T. (1972). *The Awkward Class,* Clarendon Press, Oxford.

SECTION II

GROWTH, EQUITY AND LABOUR ABSORPTION

5

GROWTH AND EQUITY IN AGRICULTURE

M.L. DANTWALA

Reviewing the state of the Indian economy, the Seventh Five Year Plan refers to some of the salient achievements of planned development. There has been a marked decline in the incidence of poverty. Based on the information available from the National Sample Survey, it is estimated that during the period 1977-78 to 1984-85 the percentage of population below the poverty level has declined from 48.1 to 36.9. Agricultural performance has been particularly impressive, specially in foodgrains. (The foodgrains stocks at the end of June 1986 were estimated around 28.6 million tonnes). The targets set under the Sixth Plan for coverage of poor families under the Integrated Rural Development Programme (IRDP) and employment generation under National Rural Employment Programme (NREP) and Rural Landless Employment Guarantee Programme (RLEGP) were met. During the Sixth Plan, the targets for elementary education enrolment and for the provision of primary and subsidiary health centres were exceeded.

Based on this optimistic picture of performance, the Seventh Plan envisages that ''The percentage of population with a consumption standard below the poverty line is expected to come down from an estimated 36.9 per cent in 1984-85 to 25.8 per cent in 1989-90," and by the end of the century to just 5 per cent. "In absolute terms, the number of poor persons below the poverty line is expected to fall from 273 million in 1984-85 to 211 million in 1989-90, the bulk of this improvement being in the rural areas.'' As for employment, ''Over the Seventh Plan, employment potential is expected to increase by 40 million standard 'person years against an increase in labour force of around 39 million persons. The bulk of growth in employment potential is in the agricultural sector, and within the sector, in subsidiary activities other than crop production." Out of the increase of 40 million standard person years, about 6.9 million person years will come from the crop sector and 11.0 million from the non-crop sector.

Growth of Agriculture

The analysis of performance and potential of the agricultural sector made below suggests that the capacity of the sector to contribute to the alleviation of poverty and unemployment is perhaps, over-estimated. Demographic pressure on land resources and environmental degeneration are the two main reasons for this caveat.

Despite wide fluctuations in agricutlrual production due to frequent natural calamities like droughts and floods, India's agricultural growth since Independence is indeed creditable. The annual rate of increase in agricultural production based on Index Numbers was 2.6 per cent between 1949-50 and 1985-86 (CMIE, 1986). Foodgrains production increased from a three-year moving average of 57.47 million tonnes in 1950-51 to 142.7 million tonnes in 1983-84, giving an annual rate of increase of 2.55 per cent. In fact, the annual compound growth rate of major foodgrains has steadily gone up from 2.16 per cent during 1950-51 to 1967-68 to 2.62 per cent during 1967-68 to 1983-84 and to 3.00 per cent during 1973-74 to 1983-84. Such sustained increases in a large country like India have been rare, historically and world over.

Demographic Pressure on Land

But all this achievement pales into insignificance when viewed in per capita terms, indicating the stress imposed by demographic pressures on the economy. Between 1951 and 1981, India's population has increased from 361 million to 685 million, and by March 1986, is expected to go up to 761 million, more than doubling within 35 years. In another 15 years, by the year 2001, India's population will hit the 1000 million mark. These facts are well known, but their implications for the national economy and particularly for the agricultural sector have not been thoroughly analysed and explained.

Data on per capita food consumption, based on the estimates of net availability of foodgrains, computed as net production (+) net imports,

(—) changes in stocks with Government provide a dismal picture. Dandekar (1986) has estimated that per capita per annum gross consumption of foodgrains has increased from an average of 181.80 kg. in 1954-58 to 185.18 kg in 1976-83, an increase of just 1.86 per cent in 30 years. The only relieving feature in this 'development' is the cessation of dependence on imports. Assuming a normal production of 150 million tonnes of foodgrains and population of 751 million in 1985, Dandekar (1986) has calculated that the net availability (allowing food seed, feed and wastage) of foodgrains from domestic production will amount to 175 kg. per capita per annum, or 480 grams per day, with a calorie value of 1632. Taking into account additions to the calories available from vegetable fats and sugar, his estimate shows that the per capita per day supply of calories in 1985 would amount to 2054, against the recommended norm of 2300 calories. Assuming further that the small deficit of 250 calories will be made up by other items of food consumption, it may be said that in 1985, India's population had a diet which, on an average, was adequate in respect of calories. But, he makes a pertinent point that "in view of the known inequality in the distribution of purchasing power in the population, the conclusion is inescapable that at least half the population lives on diet inadequate even in respect of calories."

Two points are emphasised in the above observation. When agricultural performance is measured in per capita terms, *i.e.,* with population as the denominator, much of the shine on it is wiped off. Second, unless inequality in the distribution of purchasing power is reduced, food for all will remain an unfulfilled objective. The question before us is to what extent agricultural development strategy can help to solve these problems? Obviously, the growth rate of production must not only be maintained but improved considerably. This will not be easy, but is certainly not inconceivable. A recent FAO study (1982*b*) has stated that by the year 2000, assuming that India's massive irrigation development plans materialise, and irrigation's contribution to production increases from the present level of 56 to 87 per cent, even with 'low level of inputs,' India's lands will have the capacity to support a population of 1298 million, against the assumed population of 1036 million in that year (vide CSE, 1985, pp. 157-158).

Though it is difficult to share this optimism in view of the increasing degradation of India's lands through soil erosion, waterlogging, salinity, etc., and the rapid increase in the proportion of marginal operational holdings, to which reference is made later, let us assume that the overall availability of food will not be a serious problem for India in years to come. That leaves the question of income distribution and provision of gainful employment for all, and the agricultural sector's responsibility in the matter. Before discussing this issue, let us refer to some other aspects of the country's agricultural performance during the past three decades.

Disparity in Agricultural Growth

Though the overall performance of our agriculture was satisfactory, there was considerable inter-regional inequality in growth. There is much concern over the fact that the bulk of the increase in foodgrains production has come from a relatively small area, the so-called Green Revolution belt, and consequently, as Dandekar and several other scholars have stated, "the benefits of the increase in production remained largely confined to the farmers in this area." The Seventh Plan has also stressed that "While India's agriculture has taken massive strides during three and a half decades of planning, its growth and development has not been uniform all over the country. The differential pattern and pace of development, particularly the growth of foodgrains production, has led to regional disparities" (Government of India, 1985 *a*, p. 14). The Plan document also bemoans the fact that less than 15 per cent of the area under foodgrains in the country (mostly the Green Revolution belt) contributed as much as 56 per cent of the increase in foodgrains production in the post-green resolution period (Government of India, 1985 *b*, p. 2).

Several questions, arise from these constant references to the accentuation of regional disparities in agricultural development. First, is it correct to say that the benefits of the green revolution have remained confined to the farmers in that area? Did not the whole nation benefit from the larger availability of foodgrains and its price restraining impact? Had the green revolution been stifled on ideological grounds of being 'technocratic', the poor in the country would have been the worst sufferers from foodgrains shortages and their high prices. It should also not be forgotten that the entire nation was saved the humiliation of pathetic dependence on 'foreign food'—as one ideologue put it—, thanks to the green revolution and the farmers who ushered it. Second, it is economically realistic to expect uniform growth of foodgrains production in all regions of this vast country? Agro-climatic environment varies vastly from region to region. Even the irrigation potential is not uniformly distributed over all regions. Some regions are better suited for foodgrains production than others. It would be perhaps more realistic to expect overall agricultural production to be more uniform inasmuch as regions not suited for utilizing the available foodgrains technology—particularly irrigation—could grow non-foodgrains crops more profitably, or develop their horticulture and plantations, and make up the balance. Would it make sense to push Kerala, for example, into growing foodgrains when its advantage lies in developing its plantations? It would be another matter if the lag in foodgrains production growth in some regions were attributable to the neglect of effort to exploit their full growth potential. Incidentally, information available to us on coefficient of variations in State Domestic Product per hectare of net sown area indicates that the coefficient has, in fact, declined from 41.03 in 1970-71 to 38.43 in 1977-78.*

A regionally more balanced growth of foodgrains is certainly desirable, but the pursuit of this objective should not be at the cost of pushing up the marginal cost of production of foodgrains. it would be economically more sensible to assist each region to make the optimum use of its development potential, instead of seeking product uniformity in the growth process. Even with the concentration of foodgrains production in the most favourable area, we are growing foodgrains at a cost and a price which millions in the country cannot afford, which no foreign buyer is prepared to pay, and which is therefore accumulating, perhaps rotting, in Government godowns and on open plinths. The Planning Commission should calculate the cost-benefit consequences of regionally more balanced foodgrains production. The Seventh Plan has cautiously referred to the "decline in productivity of inputs" (Government of India, 1985 *b,* p. 2). Is this a consequence of pushing production to less and less fertile lands? There is evidence of decline in fallow rotation and grazing land being converted for crop farming. A recent study of cardamom plantation in Kerala has observed that "productivity decline in cardamom has been a consequence of the large scale deforestation taking place in the Western Ghats" (CDS, 1986). The Study Report states: "The policy of the Government emphasised provision of cultivable land in the arable forest land of the State to land hungry peasants and landless labourers." The Report adds:

* *Courtesy:* S.D. Sawant; see also Dholakia (1985).

"In the wake of large scale occupation of Cardamom Hill Reserve, even areas not at all suitable for raising cardamom happened to be encroached upon" and "lands suitable for cultivation of cardamom have been converted for cultivation of seasonal crops."

Paradox of Growth

Another example of dwarfing of agriculture's performance by demographic pressure, observed by conversion of aggregates into per capita terms, is provided by Dandekar. He has pointed out that "in spite of the per capita production of almost 200 kgs., the gross per capita consumption does not seem to rise above 185 kgs. per annum." The anomaly is examined by reference to the increase in per capita Net Domestic Product and its distribution among different sectors of the population. The share of agriculture in Net Domestic Product (NDP) has declined from 58.69 in 1950-51 to 37.48 per cent in 1982-83. But the share of the population dependent on agriculture (cultivators and agricultural labourers) has not declined to the same extent. In fact, the decline during 1951 to 1981 was only one percentage point, from 67.5 to 66.5. Due to the combined effect of the decline in agriculture's share in NDP and the near stagnancy of population dependent on agriculture, the *per workers* NDP in 'agriculture, forestry, fisheries' sector declined from Rs. 1,305 in 1970-71 to Rs. 1293 in 1981 (Dandekar, 1986).

Dandekar has thus highlighted once again this paradox of growth in overall agricultural production and the declining per workers NDP in agriculture with well marshalled statistical data. The task now is to unravel the analytical significance of this phenomenon for public policy.

In this connection, our first contention is that whatever has happened to the per capita production/consumption of food or to the per worker NDP in agriculture is largely attributable to the denominator (population, labour force), rather than to the numerator (agricultural performance). In other words, the failure is due to our inability to check population growth and to sufficiently diversify the rural economy to reduce the pressure of population on agriculture.

Indian planners have consistently under-estimated the likely growth in population and over-estimated the likely shift of labour force from agricultural to non-agricultural sector. As a consequence, agriculture has been burdened with supporting a much larger population and a much larger labour force. The First Five Year Plan which envisaged that the national income could be doubled by 1971-72 and per capita income doubled by 1977-78, had assumed, *inter alia,* that population would grow at the rate of 1.25 per cent per annum during the entire period to which the projections related. The Second Plan revised this assumption upward but still assumed that the population will grow at the average rate of 1.33 per cent per annum during the decade 1961-70 and at the average rate of 1.4 per cent per annum in the decade 1971-80. As it is, during the decade ending 1971 the population grew at an average annual rate of 2.20 per cent and at the rate of 2.25 per cent in the decade ending 1981. "The Planning Commission had just put finishing touches to its Sixth Five Year Plan (1980-85), when the Registrar General of the Census announced that the country had some 13 million more people than the number estimated by the planners" (CSE, 1985).

As for the classification of labour force, the Second Plan had assumed that the total working force will increase by 19 million between 1951-1961 and by 23 million between 1961-1971, i.e., by 42 million over a period of 20 years or by 33 million during the next three Plan periods. It was stated that "If the economy develops at the sort of rates indicated in Chapter 1, it is reckoned that the proportion of labour force engaged in agricultural occupations 20 years hence (*i.e.,* by 1971) will come down to 60 per cent in place of the present proportion of 70 per cent" (Government of India, 1956). As it is, the percentage of work force engaged as cultivators and agricultural labourers was 70.1 in 1971 and 68.4 in 1981. Such serious miscalculations pertaining to critical parameters like population and labour force and policy failures which resulted from such miscalculations have not been explicitly admitted and the burden they have imposed on the agricultural sector is not fully recognised.

Impact of Rural Labour Force on Agrarian Structure

Let us now see what the unrelenting pressure of rural labour force has done to the structure of operational holdings. We believe that the most serious problem which India's agricultural economy will face in decades to come will be posed by what may be termed as marginalisation of the structure of land ownership/operation. During a span of ten years, 1970-71 to 1980-81, the number of operational holdings has increased by almost 20 million, from 70.49 to 89.39 million. During this period, the area operated increased by just 0.67 million hectares. There can be little doubt that by 1990, the number of operational holdings will reach 100 million and the operated land will remain static, or in all probability decline, as it did between 1976-77 and 1980-81.

Obviously, those 20 million additional operators (during 1971 and 1981) could not have been accommodated on the additional 0.67 million hectares of land. The additional holdings materialised from a severe and continuing process of sub-division into smaller operational holdings. Consequently, the percentage of marginal (below one hectare) holdings has steadily gone up from 50.6 in 1970-71 to 56.5 in 1980-81, and by 1990, their percentage is sure to reach 60.

At this rate, by 2001, the pressure of rural labour force on India's rural economy will be unbearable. Sundaram (1984) has estimated that in that year, with a rural population of 720 million, the rural labour force will be about 342 million, a whopping increase of 100 million in a span of 20 years from 1981 to 2001. How many of these 100 million can be accommodated on the shrinking cultivable land? What sort of land ownership/operation structure will be needed to ensure both growth and equality?

It should be admitted that there was considerable scope for orienting agricultural development strategy more purposefully towards alleviation of poverty. This scope was not fully exploited. While accentuation in regional inequality in the process of growth is perhaps unavoidable to a certain extent, accentuation of inter-class inequality within a homogeneous region could and should have been checked directly through reform of the agrarian structure and institutions. The most recent data from the 37th Round of the National Sample Survey on Assets and Liabilities of Households as on 30th June, 1981, reveals that in the rural household sector about 39 per cent of households with total assets of less than Rs. 10,000 each shared only five per cent of the value of total assets. At the other end, about eight per cent of households with assets worth Rs. 1 lakh each accounted for 46 per cent of value of total assets. Incidentally, inequality in asset ownership is even more glaring in the urban household sector, but the academicians have chosen to concentrate almost exclusively on rural inequality, perhaps because there are no easy academic solutions to the problem, such as land redistribution, in the urban sector.

Abolition of Zamindari and such other intermediary tenures was indeed a radical reform for this purpose, despite some of the loopholes.

But beyond this, performance of land reforms has been dismal. The land ceiling legislation was not sincerely implemented. Its wanton evasion was knowingly ignored. The damage done is irreversible, forcing this author to take a highly unpopular view that the land ceiling legislation has now lost much—though not all—of its potential to make a significant contribution to poverty alleviation. As for the tenancy legislation, while some States—like Gujarat and Maharashtra—utilised it to enhance social justice, a large number of other States have connived at its non-implementation. Share-cropping and concealed or fraudulent tenancy are still rampant. Given the political will, there is still considerable scope for deriving social justice from purposeful enforcement of tenancy laws. The failure in speedily implementing the less spectacular but nonetheless crucial land reforms such as updating and correcting land records and consolidation of holdings indicates disorientation of priorities. Last but not the least disturbing is the phenomenon of privatisation of "common land" to which some scholars have drawn pointed attention. Common Property Resources (CPRs) play a significant role in the life of the rural poor. Based on data from over 80 villages in 21 districts in the dry region of seven States, Jodha's study (1983) reveals that "the per year per household income derived from CPRs ranged between Rs. 530 and Rs. 830 in different States, higher than the income generated by a number of anti-poverty programmes in many areas." Regrettably, the area under CPRs and their productivity are declining. "Large scale privatisation of CPRs has taken place mostly during the last three decades. Privatisation of CPRs was done largely to help the poor... However, 49 to 86 per cent of CPRs ended up in the hands of the non-poor in different areas." Kurien (1986) has elucidated the deleterious consequences of privatisation of common property for the weaker sections. "Traditional rural society was hierarchical and inequitious, yet it was also of mutual obligations. But growing privatisation of resources and the commercialisation of economic activity have almost completely marginalised the weaker sections who find increasingly that they have to buy things which they formerly used to receive in the form of traditional claims. . . Special mention must be made of the manner in which common property resources—grazing lands, waste lands, forests, water resources—are coming to be eroded through the interrelated processes of privatisation and marketisation."

Equity

Degradation of forests and dwindling of grazing land are also hurting the poor most (CSE, 1985). The point we wish to emphasise is that while anti-poverty programmes are launched by the dozens, social and economic processes which continuously deplete the resources from which the poor derive their income and employment and make their traditional skills redundant are either not noticed or ignored and effective action to stop such processes is rarely taken. Thus, what the Government gives to the poor through its anti-poverty programmes is taken away, not too unstealthily, by the social process dominated by the rich.

In the context of the scenario depicted above, what sort of agricultural development strategy can conceivably help the agricultural sector to contribute more positively to the objective of growth with equity? It is obviously beyond our competence to suggest something quite original in this regard. At best, we can refer to some of the major issues which have sprung up from an earnest debate on alternative strategies.

The Indian planners and policy-makers have been keenly aware of the stupendous problem of growth and reconciling it with equity. The approach to the problem has maintained its facade but not the weightages of its internal composition. It is not necessary to chronicle all the shifts in approach from Plan to Plan and over the three and a half decades of development planning. Much has been written on it. The Fifth Plan was perhaps more specific on development strategy and initiated what has come to be known as a direct attack on poverty, which however was not restricted to target oriented special anti-poverty programmes. The Fifth Plan recognised that poverty abolition would entail a massive transfer of resources from the top three deciles to the bottom deciles.

In the Seventh Plan, the emphasis on asset and income redistribution is rather muted. There is hardly any search for new, imaginative and innovative approaches to reduction in assets and income inequalities. Since the poverty levels have come down significantly, according to the Planning Commission, what is now needed, in its view, is a more determined effort to push forward the current anti-poverty programmes and implement them more efficiently. The emphasis now is on science and technology for the solution of the poverty problem. To quote from the Deputy Chairman's Preface to the Seventh Plan: "Recent experience suggests that by harnessing the forces of modern science and technology, it is possible, as never before, to ensure that chronic poverty need not be the inevitable lot of a majority of humankind.'' This is what is often derisively characterised as a 'technocratic' approach to development.

There is however, one reference in the Preface which I must cite and endorse. Along with the reiteration of the need to sustain the tempo of modernization and social development, the Preface observes: ''Simultaneously, we must evolve new structures, new attitudes, a new moral code, a new work ethic, a sort of cultural revolution, if you wish, which lays emphasis on dedication and commitment. . . .'' Somewhat rhetorical, but the reference to attitudes, moral code, work ethics are very welcome and opportune. I only wish, the contents of the moral code and new ethic were spelt out in the context of development strategy whose principal objective is social justice.

The Preface refers to "the dream of Mahatma Gandhi to wipe out the tears from the eyes of each and every individual in our country". But the Mahatma also spoke of Trusteeship, exhorting the capitalists and the land-lords to renounce the rights and privileges of ownership. At the Round 'Table Conference he had proclaimed" ''When Independence comes, every title to property would be subjected to scrutiny, and confiscation ordered where necessary.'' The Plans talk constantly about poverty and its alleviation but scrupulously avoid even a mention of unmerited affluence. Do the planners believe that poverty would be eliminated without touching affluence? Is the ethics which permits poverty to be generated and sustained different from the ethics which generates affluence?

Structural Issues

This brings us to the major critique of the development strategy pursued during the planning era. Let us first put it in moral terms before it gets enmeshed in the technical jargon. Briefly it says: equity oriented policies and programmers pursued within the cast-iron inequitous economic structure—of ownership of assets—will not only be self-defeating, but may prove counter-productive, through a 'trickle up'. More simply, a direct attack on poverty without an equally direct attack on the structure, which has bred poverty and continues to do so, is an illusion at best, fraud at worst. The dilemma of 'A Moral Man in an Immoral Society' which the philosopher Neibuhr has posed!

A simulation model developed by Sinha *et al.** (1979) shows that "a spill over effect of a net injection of one rupee to the rural bottom class would result in an overall increase in income of Rs. 1.916 in the rural areas, distributed in Re. 0.213 to the bottom class, Re. 0.520 to the middle class and Rs. 1.183 to the top class (apart from an income generation of Re. 0.640 in th urban areas)" (quoted by Kurien, 1986). Another simulation model worked out by Adelman and Robinson (1978)—which I had cited in an earlier paper (Dantwala, 1983)—leads to a surprising conclusion that "most anti-poverty policies eventually help the rich and the middle income groups more than they help the poor. This is so even when, as in our experiments, the rich are taxed quite progressively to finance the programmes, the programmes are designed so that their initial impact is quite specific in favouring the lower income groups, and there is no graft, corruption, diversion, or stupidity in their execution. This triple up effect was evident in a great many different policy experiments and is difficult to avoid. Second, our experience with a wide range of policies indicates that it is much easier to make the income distribution worse than to improve it."

Summing up the findings of some of these studies and his own analysis, Kurien (1986) observes: "What strategies such as target group programmes attempt to do is to achieve through administrative interventions some *redistributions* in favour of those who do not have resource power. The interventions strategy, therefore, is an attempt to correct the structural *consequences* without altering the structural *characteristics*".

What may be called the 'Sinha-Adelman-Kurien' thesis deserves a serious debate. The thesis, to repeat is that given the structural milieu, policy interventions of the poverty alleviation type benefit the non-poor more than the poor. The essentiality of structural change cannot be questioned. it is, however, necessary to be more specific about the components of the alternate structure compatible with available resources. Second, since the existing power structure will not be interested in bringing about the desired structural change and would, in fact, do its utmost to prevent such change, who will accomplish this task and how?

The economists should not evade this question on the ground that it is for the political scientist and activists to provide the answer. Lastly, will the altered structure endure and retain its benevolent characte-

* The models in the book are based on Indian empirical data.

without a supportive change in the value system of the new power base? It will not be realistic to assume that the new economic structure will automatically generate an appropriate value system. We know that the structures have changed but the values needed to support the structure did not is an old structure-superstructure theme. If the change in ethics without the change in structure is futile, so is the change in structure without the change in ethics. We are back to the stimulating structure-superstructure debate.

Note : Reproduced from *Indian Journal of Agricultural Economics*, Vol. 42 No. 2, April-June, 1987, pp. 149-159

REFERENCES

Irma Adelman and Sherman Robinson (1978), Income Distribution Policies in Developing Countries: A Case Study of Korea, Oxford University Press, Oxford.

Centre for Development Studies (CDS), (1986), Development of Cardamom Plantation in the High Ranges of Kerala, K. Narayanan Nair *et al.*, Trivandrum, (mimeo.).

Centre for Monitoring Indian Economy (CMIE) (1986), Basic Statistics Relating to Indian Economy, Vol. I, All India, Bombay, August.

Centre for Science and Environment (CSE) (1983), The State of India's Environment 1984-85: The Second Citizens Report, New Delhi.

Dandekar, V.M., (1986), "Agriculture, Employment and Poverty", *Economic and Political Weekly*, Vol. XXI, No. 38 and 39, September 20-27.

Dantwala, M.L., (1983), "Two-Way Planning: Logic and Limitations—A Critical Review of Indian Experience", Paper prepared for the FAO Regional Officer for Aisa and the Pacific Bangkok, Thailand and subsequently reproduced in *Indian Journal of Agricultural Economics*, Vol. XXXVIII, No. 1, April-June.

Dholakia, R.H., (1983), Regional Disparity in Economic Growth in India, Himalaya Publishing House, Bombay.

Food and Agriculture Organisation (FAO) of the United Nations, 1981, (1982a), *Production Year Book*, Vol. 35, Rome.

Food and Agriculture Organisation (FAO) of the United Nations, (1982b), Potential Population Supporting Capacities of Lands in the Developing World, FAO/IIASA/UNFPA, Rome, Italy.

Government of India (1956), Second Five Year Plan, Planning Commission, New Delhi.

Government of India (1985a), Seventh Five Year Plan 1985-90, Vol. I, Planning Commission, New Delhi.

Government of India (1983b), Seventh Five Year Plan 1985-90, Vol. II, Planning Commission, New Delhi.

Jodha, N.S., (1983), "Market Forces and Erosion of Common Property Resources". Paper presented at the International Workshop on Agricultural Market in the Semi-Arid Tropics, ICRISAT Centre, Patancheru (A.P.).

Kurien, C.T., (1986), "Reconciling Growth and Social Justice Strategies Vs. Structures", in M.L. Dantwala *et al.* (Eds.) *Asian Seminar on Rural Development—The Indian*

Experience, Oxford & IBH Publishing Co. Pvt. Ltd., New Delhi.

Sinha, R. *et al.*, (1979), ***Income Distribution, Growth and Basic Needs in India***, Croom Helm, London.

Sundaram, K., (1984), "Registrar General's Population Projection, 1981-2001: An Appraisal and an Alternative Scenario", *Economic and Political Weekly*, Vol. XIX, No. 34, August 23.

6

GREEN REVOLUTION AND THE SMALL PEASANT

A Study of Income Distribution in Punjab Agriculture

G.S. BHALLA and G.K. CHADHA

The success of the Green Revolution in Punjab has attracted a great deal of attention of the researchers. That the new production technology has been widely adopted and has led to a phenomenal increase in agricultural output and income is universally recognised as a positive contribution by all scholars. Serious doubts have, however, been raised, both in India and abroad, regarding the impact of the new technology on the level of living of marginal and small farmers and landless labour households. The initial reaction of many well meaning scholars was to characterise it as a basically capital intensive technological change that by its very nature was bound to benefit only the upper strata of the peasantry and thus lead to an accentuation of tensions in the countryside. Some scholars, basing themselves on a two-point analysis of NSS data, tried to show 'convincingly' that green revolution in Punjab had not only led to accentuation of inequalities, but had also resulted in perpetuation of poverty and destitution.[1]

Many of these studies were not only based on small samples but also had a limited scope of inquiry. For example, studies dealing with production effects did not go into the question of income distribution and those concerned with the latter aspect did not generally examine the related aspect of household consumption, savings and so on. A few systematic studies on consumption in rural Punjab were based on NSS data on household expenditure. But even these studies, while giving useful information on distribution of consumer expenditure in the rural and urban areas, did not provide any meaningful data on the extent of adoption of the new technology, production efficiency and income levels by different strata of cultivating population. Taken as a whole, the studies conducted so far are basically disjointed bits of the changing agrarian reality in the Punjab. This has been a big research gap.

Under the auspices of the Indian Council of Social Science Research and the Punjab State Planning Board, a comprehensive study on income distribution and related aspects of agriculture was recently completed by the authors. The project aimed at analysing the resource structure, investment level and its composition, production, income, consumption and saving patterns of different farm size categories. Using a two-stage stratified random sampling technique, a total of 1,663 farming households belonging to six farm-size categories were selected from 180 sample villages of the state. The six categories of cultivating households, defined on the basis of net operated area, were: (1) 0.10 to 2.49 acres, (2) 2.50 to 4.99 acres; (3) 5.00 to 7.49 acres; (4) 7.50 to 12.49 acres; (5) 12.50 to 24.99 acres; (6) 25.00 acres and above. The first two categories are termed as marginal farmers' and 'small farmers', respectively. With July 1974 to June 1975 as the reference year, the actual data collection was completed in two rounds.

The main purpose of this paper is to evaluate the impact of green revolution on income distribution among cultivating households of rural Punjab. This has been derived by analysing the production structure and income levels of various farm-size categories. An attempt is also made to examine the emerging pattern of consumption with increase in income brought about by technological change in agriculture. Whereas our analysis covers all categories of cultivators in general, the main focus of the study is to bring out the relative position of marginal and small farmers in the production, income and consumption nexus.

The paper is divided into five sections. The next section deals with production structure of the various categories of cultivators and examines the relationship between farm size and productivity. The levels of farm and non-farm income accruing to each of the six farm size groups and the extent of inequalities in the distribution of income are worked out in Section III. Section IV deals with household consumption, per capita commodity expenditures and provides an idea about the extent of poverty among the cultivating households in Punjab. Finally, the main conclusions and policy implications emerging from the study are given in Section V.

II. FARM SIZE AND CROP PRODUCTION STRUCTURE

Farm Size and Cropping Patterns

Many studies conducted during the pre-green revolution period showed that there were significant differences among farm size groups in the matter of cropping patterns. The most notable among these studies is that of Krishna Bharadwaj.[2] However, since mid-sixties, the production conditions and the cropping patterns have changed a great deal, especially in areas such as Punjab.

Table 6.1, based on our 1974-75 sample data, gives a bird's eye view of the cropping patterns pursued by the six farm size groups. The Table clearly shows that the picture of mid-seventies is vastly different from the one that obtained during the mid-fifties, as described by Bharadwaj and others. Evidently, the size-specific constraints of the type suggested by Bharadwaj seem to have become non-operative. The 'distress cropping pattern' pursued by marginal, and to a lesser extent, small farmers, is nowhere in sight. The lucrativeness of a crop is a function both of its yield and relative price level. In a situation where the yield levels for different crops have experienced varying degrees of upward shift, even slightly lower relative prices have made the high yielding crops much more lucrative. Thanks to spectacular improvements in their yield levels, and the incentive of stable prices through the state policy of price support and procurement, wheat and rice have emerged as the most important commercial crops of the state. On the other hand, crops such as gram and cotton have become less important in the hierarchy of commercial crops. Table 6.1 reveals that the percentage of gross cropped area under wheat and rice remains more or less constant over different farm size categories. This indicates that the adoption of these crops is clearly size neutral. For most other crops also, the small farmers are not much different from the higher categories. Only in the case of marginal farmers, some differences are observable *vis-a-vis* higher categories. For example, the marginal farmers devote much higher percentage of their area to maize (21.5 per cent) compared with any other farm size group (whose share ranges between 7.02 per cent to 14.84 per cent). On the other hand, they devote relatively smaller area to the cultivation of cotton and sugarcane. To the extent that per acre fertiliser requirement for rice cultivation is higher compared with that for maize, marginal farmers obviously cannot go too far in substituting rice for maize, although rice cultivation promises them higher income and greater employment per acre cropped. Furthermore, rice cultivation requires ownership of assured means of irrigation like pumpset and tubewell. It is, therefore, the differences in the resource base including assured irrigation and credit availability that explains the cropping pattern differences during Kharif between marginal farmers on the one hand and the rest of the farmers on the other.

TABLE 6.1 : Cropping Pattern by Farm Size Group—Percentage of Cropped Area under Different Crops, Punjab State

Sl. No.	*Crop*	*Farm Size Groups*						
		I	*II*	*III*	*IV*	*V*	*VI*	*All Groups*
1.	Wheat	41.30	41.22	39.45	38.30	37.06	36.83	38.08
2.	Rice	10.24	12.24	10.55	10.01	8.54	11.04	9.97
3.	Maize	21.50	12.88	14.84	12.39	10.64	7.02	11.85
4.	Bajra	1.02	1.93	2.64	2.44	3.74	3.46	2.95
5.	Cotton	4.78	6.28	6.64	9.89	11.70	13.48	10.04
6.	Sugarcane	1.37	2.42	2.15	2.19	2.13	2.15	2.14
7.	Gram	1.02	1.29	2.54	3.25	4.35	4.84	3.55
8.	Fodder	13.65	17.23	15.33	13.96	12.62	10.99	13.59
9.	Others	5.12	4.51	5.86	7.57	9.23	10.19	7.83
10.	All Crops	100.00	100.00	100.00	100.00	100.00	100.00	100.00

These differences notwithstanding, it appears that small and marginal farmers of the Punjab have adjusted their cropping patterns in full response to the opportunities available to them. We have seen above that all crops are grown by farms of all sizes, in spite of the existence of small inter-group differences in the percentage of cropped area under individual crops. The degree of commercial orientation is evident as much and operates as freely on marginal and small farms as on medium and large ones. Neither is small land area a special disadvantage, nor a large land area an extra advantage in the choice of crop combination. Consequently, a more or less uniform set of priorities is observed by farms of different sizes. For example, all farm size groups accord first priority to wheat, second to fodder (combined), and, third and fourth shared between maize and rice.

It may be relevant to point out, that in the matter of cropping patterns and priorities, the marginal and small farmers of Punjab, particularly the latter, stand close to medium and large farmers, mainly on the strength of their equally strong irrigation base.[3] This is probably a peculiar feature of Punjab agriculture and may not have its counterpart in most areas outside Punjab. All the same, it has vital implications for those states of India which are attempting technological transformation of their agriculture on the lines of Punjab. The vast majority of their marginal and small farms cannot switch over to more remunerative crop varieties, as long as irrigation remains a bottleneck with them.

Farm Size and Input Use

Table 6.2 sheds light on the composition of material costs, i.e., expenditure on seeds, manure and fertilisers, maintenance of bullocks, diesel and electricity, repair and maintenance of implements, hiring of implements/bullocks and on other items by farm size groups. It is evident that expenditure on material inputs per acre of cropped area is inversely related to farm size. On marginal and small farms, a lion's share of material costs is on account of expenditure on draught cattle. This constitutes as high as 48.63 per cent of total material cost by the marginal farmers. There are presumably technological compulsions for small farms to maintain this costly input and consequently suffer scale indivisibilities. Further, as the farm size increases, the share of bullock expenditure drops steeply, while the share of expenditure on diesel and electricity, repairs and maintenance of implements/machinery, etc. goes on increasing. The share of bullock expenditure declines from 48.63 per cent for marginal farmers to only 13.15 per cent for the highest category. On the other hand, expenditure together on diesel, electricity and repair and maintenance of implements increases sharply from 7.85 per cent of total material cost for marginal farmers to 25.28 for the highest category. The marginal and small farmers are spending slightly less on per acre application of manures and fertilisers. It is worth remembering that there was a steep hike in fertiliser prices in the survey year (1974-75) and fertiliser consumption in the state dropped by a significant margin in 1974-75 over 1973-74.[4] In such a situation, the marginal and small farmers are bound to be adversely affected, particularly because of their limited resource base. Nevertheless, in spite of this sudden change in fertiliser prices, the smaller holdings in the Punjab compared fairly well in the use of this most crucial input.[5]

The overall impression that clearly emerges from Table 6.2 is that

TABLE 6.2 : Composition of Material Cost by Fame Size Group, Punjab State

Item	Farm Size Groups													
	I		*II*		*III*		*IV*		*V*		*VI*		*All*	
	Total	*Per Acre*	*Total*	*Per Acre*	*Total*	*Per Acre*	*Total*	*Per Acre*	*Total*	*Per Acre*	*Total*	*Per Acre*	*Total*	*Per Acre*
Exp. on diesel and electricity	42.39 (3.95)	14.30	190.75 (8.06)	29.48	308.69 (9.34)	30.09	542.51 (11.73)	33.95	1007.40 (13.73)	38.29	2883.34 (17.77)	49.67	543.22 (12.18)	36.26
Exp. on draught cattle	521.64 (46.63)	175.93	1087.85 (45.96)	168.20	1297.17 (39.23)	126.44	1404.69 (30.37)	87.90	1606.22 (21.89)	61.05	1763.96 (13.15)	36.76	1295.67 (29.05)	86.49
Exp. on repair and maintenance of implement	41.88 (3.90)	14.12	99.25 (4.19)	15.34	144.74 (4.38)	14.11	223.36 (4.83)	13.98	372.80 (5.08)	14.17	1007.77 (7.51)	21.00	227.42 (5.10)	15.18
Exp. on marketing and transport	5.41 (0.50)	1.82	18.38 (0.78)	2.84	38.57 (1.17)	3.76	54.36 (1.18)	3.40	120.90 (1.65)	4.60	308.24 (2.30)	6.42	63.05 (1.41)	4.21
Hire charges paid out	43.96 (4.10)	14.83	58.18 (2.46)	8.99	69.80 (2.11)	6.80	91.50 (1.98)	5.73	102.22 (1.39)	3.89	108.20 (0.81)	2.25	78.44 (1.76)	5.24
Exp. on seeds	104.70 (9.76)	35.31	213.74 (9.03)	33.03	360.47 (10.90)	35.14	561.76 (12.15)	35.15	932.74 (12.71)	35.45	1836.49 (13.69)	38.27	532.03 (11.93)	35.51
Exp. on manures and fertiliser	275.22 (25.65)	92.82	633.91 (26.78)	97.96	1007.34 (30.47)	98.19	1598.49 (34.56)	100.02	2958.80 (40.31)	112.43	5386..63 (40.16)	112.23	1576.53 (35.35)	105.24
Other exp. (water and pesticides)	37.62 (3.51)	12.69	64.85 (2.74)	10.02	79.58 (2.41)	7.76	147.99 (3.20)	9.26	237.58 (3.24)	9.03	618.91 (4.61)	12.90	143.44 (3.22)	9.58
Total	1072.82 (100.00)	361.83	2366.91 (100.00)	365.75	3306.36 (100.00)	351.51	4624.66 (100.00)	289.38	7738.66 (100.00)	278.93	13413.54 (100.00)	279.55	4459.80 (100.00)	297.71

Exp. refers to Expenditure

smaller farms depend relatively more on traditional capital inputs (e.g. draught cattle), although they try to supplement their own stock of productive assets by hiring in machine services. That is why, per acre expenditure on machine hire charges is much higher on marginal and small farms (item 5. Table 6.2).

Farm Size and Cropwise Use of Manures and Fertilisers

We have seen above the relative position of the six farm size groups as regards manure and fertiliser use per acre of gross cropped area. it might also be useful to see per acre fertiliser application for individual crops. Table 6.3 provides some information in this regard. Since fertiliser application is generally confined to irrigated cropped area (in any case, in Punjab unirrigated area is a very small percentage of total cropped area), only irrigated areas under individual crops are considered. However, for the total of all crops, both irrigated and unirrigated areas are included. To formulate better ideas about input use of individual crops, the package of bio-chemical inputs (comprising manures, fertilisers, seeds, and pesticides) is also included in the Table in addition to the total of manures and fertilisers alone.

A few things emerge clearly from Table 6.3. Firstly, the inter-group differences are much smaller with respect to the total of bio-chemical inputs than manures and fertilisers alone. Secondly, the inter-group differences are operating relatively more for paddy cultivation than for other crops. This is explained possibly by the fact that the rice revolution commenced only in early seventies and, as in the case of the wheat revolution of the sixties, was more or less confined to the large farmers to begin with. Till the time that the rice revolution gets diffused among most of the rice cultivators including marginal and small farmers, farm size-wise differences in input use, including per acre use of fertilisers, are bound to exit. The relatively mild inter-group differences observed in the case of wheat lend support to this reasoning.

Farm Size and Yield Rates of Individual Crops

The analysis in the previous section makes it fairly evident that small and marginal farmers of Punjab, although bound to the technological compulsions specific to their limited land area and other resource constraints, are doing everything possible to follow the medium and

TABLE 6.3 : Input use by Farm Size Group, Punjab State

(Rs. per Cropped Area)

Farm Size Group	Crop Type				
	Wheat (Irr)	*Paddy (Irr)*	*Maize (Irr)*	*Cotton (Irr)*	*All Crops (Irr + Unirr)*
Manure and Fertiliser					
I	131.59	126.30	115.00	83.66	92.82
II	134.26	126.00	135.48	117.43	97.95
III	140.01	133.46	127.50	96.39	98.19
IV	141.31	152.80	115.56	96.76	100.02
V	157.63	169.75	141.22	102.54	112.46
VI	147.28	183.95	99.51	110.01	112.26
All Farms	146.57	156.43	126.34	102.43	105.24
Bio-Chemical Inputs					
I	174.58	142.40	128.56	114.78	129.08
II	176.95	145.65	150.38	152.41	132.75
III	182.22	148.93	141.91	125.31	134.90
IV	181.55	176.43	129.07	132.51	138.66
V	199.19	190.28	155.53	178.65	152.60
VI	189.60	214.96	113.93	162.86	157.43
All Farms	188.13	178.49	140.49	142.17	144.48

$$\text{Log } Y_1 = 0.231 + 0.049^* \text{ Log } X$$
$$(4.06)$$

$$\text{Log } Y_2 = 0.308 + 0.049^* \text{ Log } X$$
$$(4.20)$$

where Y_1 is value of manure and fertilisers used per acre of gross cropped area; $Y2$ is value of all bio-chemical inputs used per acre of gross cropped area; Figures in parentheses are 't' values; and * denotes significance at 0.01 level.

the large farmers in the use of new technological inputs. Although some differences are noticeable as regards the level of per acre application of resource, inputs, most pronouncedly for paddy cultivation, yet in a general sense, the resource use pattern of different farm size groups is strikingly similar. To the extent that yield rates for individual crops depend directly on the quantum of resource use, we have reasons to believe that yield rates should not generally differ significantly by farm size groups, except in the case of crops such as rice, Table 6.4 lends considerable support to our thinking. The regression equation in Table 6.4 are based on double-log functions in order to absorb the element of non-linearity that generally accompanies the use of data on per acre basis.

For all crops except paddy, the yield rates are generally invariant

TABLE 6.4 : Yield Rate per Acre of Cropped Area by Farm Size Group, Punjab State

Sl. No.	Crop	*Farm Size Groups* I	II	III	IV	V	VI	All Groups	*Regression Equation between Yield Rate (Y) and Farm Size (X) (y = Log Y; x = Log X)*
1.	Wheat	1000.91	100.91	1071.31	1096.33	1096.53	1071.31	1067.65	y = 2.952 + 0.029x (1.530)
2.	Rice	934.50	995.07	1070.48	1131.48	1196.06	1272.31	1140.63	y = 2.788 + 0.172*x (3.094)
3.	Maize	707.89	798.30	624.14	661.84	665.91	554.71	646.82	y = 2.685 – 0.012x (–0.211)
4.	Cotton	824.71	989.64	910.82	947.09	982.54	982.89	966.91	y = 2.739 + 0.122*x (2.015)
5.	Sugarcane	1815.71	1696.07	2026.23	1992.83	1906.02	1826.28	1934.81	y = 3.157 + 0.042x (0.551)
6.	Gram	590.00	567.63	644.04	729.83	667.57	478.26	633.55	y = 2.639 + 0.063x (0.525)
7.	Fodder	800.00	803.75	769.62	768.71	715.19	597.61	735.21	y = 2.868 – 0.043x (–1.456)
8.	All Crops	891.29	887.92	905.22	942.52	944.93	940.63	931.38	y = 0.948 + 0.032*x (2.430)

Note : The figures in parentheses are 't' values.

* Denotes significance at 0.01 level.

along the farm size continuum; for paddy, the yield rates go on increasing as the farm size increases.[6] We noticed earlier (Table 6.3) that for paddy cultivation, the per acre application of manure and fertiliser, or of bio-chemical inputs as a whole, rises almost systematically along the farm size ladder. There is possibly an interesting explanation for this phenomenon. The small and marginal farmers tend to catch up with higher size groups in adopting new rice technology, as they had been doing for wheat during the past few years. This involves dispersal of their limited resources, over both *Kharif* and *Rabi* cropping seasons. This naturally weakens the relative position of small and marginal farmers primarily because it is not the same thing to compete for two major crops (i.e. wheat and rice) as for one major crop alone (wheat). It is difficult to conjecture whether the relative disadvantage of small and marginal farmers will disappear in due course; perhaps yes if the per acre discrepancies in fertiliser use are remedied further through institutional measures. The village co-operative societies in the state have a difficult task ahead of them, if the diffusion of the emerging rice technology has to be sustained on a scale that occurred earlier for wheat.

Farm Size, Overall Yield Rates and Productivity Per Acre

We have seen above the picture about yield rate variations of individual crops along farm size groups. Let us now see the picture aggregated over all crops. The following three equations throw up important results:

$$\text{Log } Y_1 = 0.948 + \underset{(2.43)}{0.032^*} \text{ Log } X \qquad \ldots (1)$$

$$\text{Log } Y_2 = 2.301 - \underset{(-10.99)}{0.084^*} \text{ Log } X \qquad \ldots (2)$$

$$\text{Log } Y_2 = 3.188 - \underset{(-2.44)}{0.046^*} \text{ Log } X \qquad \ldots (3)$$

Where Y_1 is overall yield rate (value of crop output per acre of *gross cropped area);* Y_2 is cropping intensity (gross cropped area ÷ net sown area) Y_3 is productivity per acre (value of gross output per acre of *net operated area);* X is farm size (in acres); figures in parentheses are 't' values, and * denotes significance at 0.01 level.

Equation (1) shows that small farmers have lost their traditional edge of higher overall yield rate presumably because of the recent introduction of new rice technology in which the small farmers were not yet on par with the higher categories. But then, the small farmers try to counterbalance their relative disadvantage of lower overall yield rate (presumably caused by lower yield rate for paddy) by achieving higher cropping intensity (Equation 2) primarily through more intensive employment of their family labour input. To the extent that (i) the irrigation base of marginal and small farmers compares very well with that of higher categories, (ii) small farmers enjoy a much more flexible and assured supply of labour all the year round, (iii) labour is a substitute for some types of capital over a wide range of production possibility curve, and (iv) marked differences in cropping patterns do not exist along the farm size continuum, the prospects of realising higher cropping intensity are relatively brighter for small farmers. Moreover, given the exigency of low land-man ratio, the ultimate limit for extending area under plough on such farms is determined not by labour and management bottlenecks (as in the case of big farmers) as by the availability of land itself.

The higher cropping intensity on smaller farms reveals its effect in terms of the inverse relationship between farm size (X) and productivity per acre of net operated area (Y_3) as equation (3) shows clearly. In other words, the smaller farmers are able to produce higher output for every acre of area held although they suffer slight setback in yield rates for some crops, notably rice, which are currently at the threshold of a new technological break-through. It is, however, notable that the inverse relationship between farm size and productivity does not hold for the central Punjab plains while it holds true for other parts of the state. This fact becomes clear when separate regressions are run between productivity per acre and farm size for the three agro-climatic regions of the state. It is a major agrarian change in some parts of Punjab that in spite of higher cropping intensity, the marginal and small farmers are unable to obtain a higher output per acre, as perhaps they did in the pre-green revolution era. The implication of this result is obvious. As capital intensity of agricultural production increases more and more in the future, the smaller holdings as individual units of production will be thrown to a position of relative disadvantage. A strong institutional support will be greatly needed to save their production base from such an erosion.

In sum, it is fairly evident that small and marginal farmers of Punjab, although bound by technological as well as pecuniary compulsions specific to their limited land area, are doing everything possible to follow the medium and large farmers in production technology. The outcome is that they compare very well with farms of larger sizes as regards output per acre.

III. HOUSEHOLD INCOME, CONSUMPTION AND SAVINGS

The income of a typical cultivating household is derived from both farm and non-farm sources. In general, the household derives its major income from cultivation, provided its land base is adequate. Farm business income of a cultivating household can be calculated by deducting all material costs (whether purchased or home produced) and paid out labour costs from the value of gross crop output. The value added in the production process can be derived from value of gross crop output by simply deducting from it the total cost on material (non-factor) inputs used in the production process. In the following calculations, no provision has been made for depreciation; hence all income concepts are in gross terms.[7]

The sources of non-farm income can be varied. In Punjab, dairying constitutes the most important source of non-farm income. Other sources of non-farm income reported by the sample households are income from poultry, wage employment inside and outside agriculture, hiring out of farm assets, income from small household enterprises and transfer payments. It is clear from Table 6.5 that non-farm income constitutes about one-fifth of total household income for cultivating households in general. For marginal farmers, it is about 65 per cent of total household income. In per capita terms also the marginal farmers earn the highest from non-farm activities, compared with all other categories.

As a result of these supplementary incomes, the group-wise gaps in per capita total household income are much smaller than in farm business income alone. For example, the ratio between the higher and the lowest categories is as high as 20.09 as regards per capita area operated, 11.25 as regards per capita farm business income, and as low as 4.54 in the case of per capita total household income. The supplementary incomes are thus acting as an important instrument in reducing the income inequalities arising out of very disparate land distribution. However, in spite of their additional efforts, the total household income of marginal and small farmers falls short of their consumption expenditure so that savings turn out to be negative in their

TABLE 6.5 : Household Income and Savings, Punjab State

Sl. No.	Item	Farm Size Groups I	II	III	IV	V	VI	All Group
1.	Total farm output	2647.04	5744.87	9207.54	15061.52	24861.00	45131.59	13952.13
2.	Material costs	1072.82	2366.91	3306.37	4624.66	7338.66	13413.54	4459.80
3.	Gross value added	1574.32	3377.96	5981.17	10436.86	17522.34	31718.05	9492.33
4.	Paid-out labour costs	104.53	374.33	673.14	1458.93	2927.86	6108.13	1422.10
5.	Total costs	1415.53	3172.67	4648.33	6978.12	11275.39	20848.51	6620.98
6.	Farm Business income							
	(a) Per household	1231.61	2572.19	4619.20	8083.40	13585.61	24283.08	7331.16
	(b) Per acre area held	754.50	721.38	776.24	846.41	825.73	740.40	800.59
	(c) Per capita per annum	212.87	421.59	689.38	1044.92	1625.44	2395.94	1015.04
7.	Non-farm income							
	(a) Per household	2271.04	1505.45	1322.25	2125.09	2644.24	3576.03	1981.82
	(b) Per capita	392.53	246.53	197.34	274.70	316.37	352.84	274.40
8.	Total household income							
	(a) Per family	3502.65	4077.64	5941.45	10208.51	16229.85	27859.11	9313.37
	(b) Per capita	605.40	658.33	886.72	1319.62	1941.82	2748.78	1289.50
9.	Percentage of households income from							
	(a) Cultivation	35.16	63.08	77.75	79.18	83.71	87.16	78.72
	(b) Non-farm activities	64.84	36.92	22.25	20.42	16.29	12.84	21.28
10.	Annual consumption expenditure							
	(a) Per family	4201.88	4872.41	5766.06	7212.67	9030.00	12887.19	6740.29
	(b) Per capita	726.25	793.60	860.54	932.36	1080.39	1271.54	933.34
11.	Household savings							
	(a) Per family	–699.24	–794.78	175.39	2995.83	7199.85	14971.92	2573.08
	(b) Per capita	–120.86	–130.27	26.18	387.26	861.42	1477.23	356.26
12.	Gross investment in farm assets							
	(a) Per household	97.19	134.88	314.80	762.44	1118.44	2784.22	631.00
	(b) Per acre operated	59.53	37.83	52.77	79.83	67.98	84.89	68.91
13.	Gross investment in house construction (per household)	2.00	0.55	4.46	3.42	26.77	16.22	7.91
14	Index of per capita income**	22.02	24.32	32.26	48.01	70.64	100.00	—
15	Index of per capita expenditure**	57.12	62.81	67.68	73.33	84.97	100.00	—

**Group VI: 100.00

case. To some extent, the deficit might be explained by the general tendency of farmers to overstate consumption and understate their income. The figures for household income are underestimated for all farm size groups, but relatively more for marginal and small farmers. Again, since our data do not take account of capital formation through sheer labour power of cultivating households, total incomes and hence savings are under estimated. On account of all these factors, the real deficit can be taken to be smaller than what appears in Table 6.5. it must, however, be stressed that the deficit is perhaps a real phenomenon among marginal and small farmers, its magnitude notwithstanding.

That the marginal and small farmers are incurring substantial deficits even after the green revolution has taken deep roots in the state is truly disconcerting. However, the mean values should not be interpreted to imply that each of the marginal and small farmers is running into a deficit. The results show that because of the average position of supplementary sources of income, and the capacity of each farm family to make use of these opportunities, the income-consumption gap varies from household to household. There might be some households among the marginal and small farm categories which just break even or earn a meagre surplus while many others incur large deficits. On the average, the two categories show considerable deficits. On the other hand, in spite of higher per capita consumption expenditure, middle and large farmers are recording substantial surpluses. A part of these surpluses goes into investment.

Table 6.5 gives data also on savings and investment undertaken by various categories of cultivating households. Some important results need to be highlighted. Firstly, it is interesting to note that despite their deficit, the marginal and small farmers continue to undertake a fair amount of investment. In fact, per acre investment undertaken by the marginal and small farmers compares very favourably with the bigger farmers. It seems that, like many other production activities, the marginal and small farmers are trying to keep pace with the bigger farmers in the matter of undertaking investment also. Since this is happening despite their deficit on current income account, the investment expenditure is being financed presumably through borrowings. The second major conclusion is that the bigger categories are using only a fraction of their surplus for capital formation in agriculture itself. There could be several reasons for it. Perhaps they have already built sufficient stock of capital. Or perhaps they are finding alternative avenues for disposing of their surplus. One does not find residential construction to be very popular either. The moot question that remains to be answered is: what happens to the huge surpluses that middle and larger farmers are able to record? Unfortunately, we are not in a position to answer it with the help of our survey data.

A note of caution is in order while interpreting the results on investment. Like data on other sources of income, excepting income from animal husbandry, the number of positive respondents in the case of investment is not very large so that the statistical reliability of the mean values used by us is not very high. The results, therefore, at best give a qualitative picture of the pattern of investment by cultivating households.

Sources of Non-Farm Income

We have already noticed the extreme importance of non-farm income to the household economy of marginal farmers. Table 6.5 further shows that the share of non-farm income declines steeply as we move towards higher farm size groups. Given the extreme importance of non-farm activities to marginal and small farmers, it is essential to identify the sources which provide supplementary income employment opportunities to such farmers. Table 6.6 provides relevant information in this regard.

However, before analysing these results, it is important to note the limitation of data on income from other sources. There are only a few activities—dairying being the outstanding example—where a very large percentage of households in each category is reporting some income. For most other sources of non-farm income, the number of reporting households being very small, the inter-group coefficients of variation are very large. Hence the statistical validity of the mean values in this case is quite low. This limitation must be kept in mind while interpreting the results of this section.

It is evident from Table 6.6 that, on an average, marginal and small farmers derive some income from each of the non-farm activities. In fact, all avenues of outside employment are welcome to them thanks to a very low level of employment and income on their own tiny farms. However, all non-farm activities are not equally important. Nevertheless, for most of the cultivating households, a major share of non-farm income is derived from dairying. The only exception is marginal farmers who obtain the highest share of non-farm income from pensions, salaries, remittances, etc. and so on.

TABLE 6.6 : Income from Non-Farm Activities, Punjab State

Sl. No.	*Item*	*Farm Size Groups*						
		I	*II*	*III*	*IV*	*V*	*VI*	*All Groups*
1.	Hiring out assets*	127.14 (5.60)	115.71 (7.69)	54.50 (4.12)	169.52 (7.98)	254.96 (9.64)	769.45 (21.52)	170.08 (8.58)
2.	Pension, remittances etc.	855.43 (37.67)	450.13 (29.90)	289.31 (21.88)	392.22 (18.46)	408.08 (15.43)	309.73 (8.66)	416.85 (21.03)
3.	Agriculture wage employment	279.68 (12.32)	78.74 (5.23)	65.20 (4.93)	9.14 (0.43)	0.00 (0.00)	0.00 (0.00)	56.67 (2.86)
4.	Non-agriculture wage employ-ment	281.65 (12.40)	206.02 (13.68)	92.23 (6.98)	140.08 (6.59)	88.53 (3.35)	378.36 (10.58)	154.30 (7.79)
5.	Household enter-prises**	165.51 (7.29)	29.45 (1.96)	37.43 (2.83)	102.10 (4.80)	6.58 (0.25)	0.00 (0.00)	55.14 (2.78)
6.	Net income from dairying and poultry	561.83 (24.72)	625.40 (41.54)	783.58 (59.26)	1312.03 (61.74)	1886.09 (71.33)	2118.49 (59.24)	1128.78 (56.96)
7.	Total	2271.04 (100.00)	1505.45 (100.00)	1322.25 (100.00)	2125.09 (100.00)	2644.24 (100.00)	3576.03 (100.00)	1981.82 (100.00)

* Includes hiring out of combine harvestor, tractor, tubewells/pumpset, thresher, draught animal and land on rent.

** Includes activities such as cottage industries.

Note: Figures in parentheses are percentages of total non-farm income.

In general, all non-farm activities are not accessible to marginal and small farmers with equal ease. For example, although dairying and poultry possess very high potential for additional employment and income for marginal and small farmers, yet initial investment required for such ventures may stand in the way. On the other hand, the availability of excess family labour and farm fodder and the prospect of steady income from the sale of milk to government or private dairies along with availability of loan from MFALA and SFDA, have provided enough incentive for most of the marginal and small farmers to keep and rear milch cattle. Other non-farm activities such as wage employment inside and outside agriculture, salaried jobs and other tertiary activities in nearby market centres and/or urban areas are almost investment free. The marginal and small farmers have greater propensity to pursue the latter types of non-farm activities as long as work on their own tiny farms does not suffer. It is interesting to point out that wage employment outside agriculture is preferred to wage employment inside agriculture, much more so by small farmers. The socio-cultural milieu in rural Punjab forbids the cultivating households from seeking wage employment inside agriculture even though such opportunities are available at their very door step. Household enterprises are yet another source of non-farm income to marginal and small farmers; farmers operating 10 acres or more do not seem to pursue such enterprises at all. At the other extreme, income from hiring out of farm assets constitutes the most important supplementary source of income for the bigger cultivators. In brief it is obvious that whereas the marginal and small farmers are trying to make maximum use of their surplus family labour, the large farmers try to optimise the use of their capital assets.

Distribution of Farm and Non-Farm Income

We now discuss the distribution of farm and non-farm income among the various strata of cultivating households. For this purpose, the households are re-arranged into decile groups, in ascending order of net operated area. Table 6.7 gives summary information on various sources of income for each decile group. To comprehend the correspondence between the decile arrangement and the farm size, the average area operated by each decile group is also given in Table 6.7.

TABLE 6.7 : Distribution of Farm and Non-Farm Incomes by Decile Group, Punjab

				Per cent Share of Income from				
Sl. No.	*Decile Group*	*Average Area Ope-rated (Acres)*	*Per cent Share in Net Ope-rated Area*	*Crop Pro-duction*	*Dairying and Poultry*	*Other Non-farm Activities*	*Total Non-farm Activities*	*Total House-hold Income*
1.	0-10	1.77	1.93	1.85	5.37	18.42	10.49	3.80
2.	10-20	3.23	3.52	3.04	5.42	12.24	8.84	4.17
3.	20-30	4.34	4.73	4.44	4.53	7.44	6.29	4.83
4.	30-40	5.40	5.88	5.70	5.39	6.87	6.03	5.77
5.	40-50	6.42	7.00	7.02	7.90	5.98	7.07	7.03
6.	50-60	7.73	8.42	8.45	10.61	5.23	8.30	8.42
7.	60-70	9.56	10.42	11.16	10.31	13.12	11.52	11.24
8.	70-80	11.75	12.80	13.19	15.54	8.18	12.37	13.02
9.	80-90	15.14	16.51	17.10	15.29	11.07	13.48	16.33
10.	90-100	25.93	28.78	28.06	18.75	11.44	15.60	25.41
11.	0-100	9.16	100.00	100.00	100.00	100.00	100.00	100.00
Gini's Concentration Co-efficient			0.381	0.394	0.256	0.048	0.125	0.336

As expected, there is a striking parallel between the decile distribution net operated area and income from crop production. In fact, the Gini's Concentration Co-efficients for the two are almost identical. Given a large inequality in resource base, one would expect the small farmers to derive a disproportionately low share in dairying income. The facts are, however, the other way round. The share of dairying income received by the smaller farmers is much higher than their share of net operated area. For example, 40 per cent of cultivators at the lowest rung of the ladder with 16 per cent of total operated area account for 21 per cent of total income from dairying. This indicates that inequalities in income from dairying are much milder than inequalities in the distribution of land and income therefrom. To some extent, this bears testimony to the effective functioning of Small Farmers Development Agency and Marginal Farmers and Agricultural Labourers' Agency. Through the lending operations of these agencies, the household economy of marginal and small farmers is reported to have improved significantly in many areas.

Income accruing together from all other non-farm activities (wage employment, hiring out of assets including land, small household enterprises, pension/salaries/remittances) shows a very interesting inter-decile distributional pattern. Whereas the bottom 20 per cent of the cultivating households obtain nearly 30 per cent of this income, the top 20 per cent get about 22 per cent. It is only the intermediate deciles that obtain a lesser share of this income. The income share of households at the bottom is made up largely of wage employment and salaries/remittances, whereas the share of those at the top comprises mainly of income from hiring out of assets. It is notable that the overall inter-decile distribution for this source of income is fairly close to the egalitarian norm. Finally, considering the total of the non-farm income, we find that its distribution shows a fairly high degree of equitable sharing among the decile groups especially in comparison with the sharing of crop income. This is perhaps a very happy situation for the lower strata of cultivating households and gives credence to the view that the modernisation of agriculture in Punjab has created many ancillary non-farm activities providing supplementary employment and income to marginal and small farmers. There is, however, tremendous scope for expanding the non-farm activities further with proper institutional and financial support, thus providing more productive employment to the weaker sections of peasantry.

IV. FARM SIZE AND HOUSEHOLD CONSUMPTION

In preceding sections of this paper, we have examined the production structure and the level of household income of the six farm size groups. We now analyse their consumption patterns. More specifically the aim is:

(i) To estimate commodity-wise per capita expenditure in respect of each of the six farm size groups, and to examine if significant variations exist across these groups;

(ii) To analyse the extent of inequalities in commodity expenditures and total per capita expenditure among the cultivating households;

(iii) To estimate expenditure elasticities for different commodities and work out their implications in terms of income levels and consumption standards; and

(iv) To work out roughly the percentage of cultivating households living below poverty line defined in terms of minimum per capita expenditure.

Before we proceed with our analysis, it is essential to point out some limitations of our consumption data. Firstly, we could not elicit from the respondent households separate expenditure figures for education, health, transport, etc, although these items are assuming increasing importance in the context of green revolution and have a bearing on the future earnings and consumption levels of rural households. These items, *inter alia,* are clubbed together under 'miscellaneous goods' and only a rough idea about their respective importance to different strata of cultivating households can be had by examining the expenditure under this heading. Secondly, there is a general tendency with the farmers to overstate their consumption of some items, specially of milk. Accordingly, for some commodities, the per capita expenditure figures are perhaps overestimated. Thirdly, while nearly all the households consume commodities like cereals, pulses, milk and milk products, edible oils, sugar, salt and spices, fuel and light, etc, the number of households reporting expenditure on meat, egg and fish, foot-wear, clothing, other durable goods, etc. constitute only a small percentage of total households in each farm size category. The mean expenditure figures for these commodities should at best be understood as indicative of qualitative contrasts.

Per Capita Commodity Expenditure

Table 6.8 gives per capita expenditure on 15 items of consumption along with their percentage share in total per capita expenditure of various size categories.

The per capita expenditure on cereals, salt and spices, and, edible oils, is more or less invariant among the six farm size groups. Over a wide range of farm size, the per capita expenditure on pulses, vegetables and fruit, and, fuel and light, is also invariant except that a relatively higher figure is noticeable for households operating 25.00 acres or more. Further, the per capita expenditure on milk and milk products, meat, egg and fish, sugar and gur, and beverages among the food items, seems generally to increase with the farm size. These results need a cautious interpretation and some clarification. The relatively inferior position of marginal and small farmers in the consumption of superior food items such as meat, egg and fish, may be explained in terms of high prices of these items in relation to their income. However, their per capita consumption of milk is much higher than their reported consumption of other superior foods, although in this case also, the marginal and small farmers incur much smaller absolute per capita expenditure when compared with the large farmers. The relatively higher expenditure on the home produced milk is due largely to the general tendency of all farmers in Punjab, to avoid reporting the sale of milk. This is evident on two counts. Firstly, the cash component of total expenditure on milk and milk products is very small by farm size groups. Secondly, our data show very little gap between milk produced and milk consumed by cultivating households of all sizes. Correspondingly, the per capita expenditure figures on milk and milk products are over-estimated for farms of all sizes. In relative terms, despite their higher propensity to consume milk, the per capita expenditure figures for higher farm size categories are bound to be relatively more over-estimated since in per capita terms they produce a much larger quantity of milk at home.

The position regarding per capita expenditure on sugar and gur, and beverages can possibly be explained in terms of changing composition of these items as we move higher on the farm size ladder. It is quite possible that marginal and small farmers have a larger component of 'raw sugar' and/or 'gur' which are relatively cheaper but perhaps nutritionally superior. Similarly, such farm families cannot incur much

TABLE 6.8 : Per Capita Annual Consumption Expenditure Commodity-wise, Punjab State

(Rs.)

Sl. No.	*Item*	*Farm Size Group* I	II	III	IV	V	VI	All Farms
1.	Cereals	177.50 (24.43)	192.00 (24.06)	189.49 (22.02)	194.91 (20.30)	205.33 (19.01)	207.71 (16.34)	195.09 (20.91)
2.	Pulses	22.20 (3.14)	28.10 (2.93)	25.50 (2.96)	26.46 (2.84)	31.66 (2.93)	32.40 (2.55)	72.30 (2.93)
3.	Milk and its products	218.22 (30.04)	248.62 (25.95)	275.58 (3.20)	312.87 (33.56)	369.03 (34.16)	388.89 (30.54)	304.27 (32.60)
4.	Edible oils	22.64 (3.12)	20.49 (2.14)	21.24 (2.47)	17.99 (1.93)	20.25 (1.87)	18.74 (1.47)	19.98 (2.14)
5.	Meat, egg, fish etc.	3.87 (0.53)	5.59 (0.58)	6.06 (0.70)	7.58 (0.81)	12.72 (1.18)	27.57 (2.17)	9.01 (0.97)
6.	Vegetable, Fruits etc.	30.41 (4.19)	30.30 (3.16)	31.43 (3.65)	33.79 (3.62)	38.42 (3.56)	50.66 (3.98)	34.51 (3.70)
7.	Sugarcane, gur, etc.	59.36 (8.17)	64.97 (6.78)	66.76 (7.76)	69.68 (7.47)	79.03 (7.31)	84.22 (6.62)	70.47 (7.55)
8.	Salt and spices	15.04 (2.07)	14.45 (1.51)	14.95 (1.74)	14.50 (1.56)	14.72 (1.36)	15.58 (1.23)	14.75 (1.58)
9.	Beverages	27.27 (3.75)	30.74 (3.21)	34.60 (4.02)	34.43 (3.69)	38.40 (3.55)	71.44 (5.62)	36.54 (3.92)
10.	Fuel and light	40.47 (5.57)	45.55 (4.75)	44.40 (5.16)	45.50 (4.88)	46.27 (4.28)	55.20 (4.34)	45.69 (4.90)
11.	Clothing	44.09 (6.07)	46.59 (4.86)	64.07 (7.44)	74.90 (8.03)	104.47 (9.67)	156.78 (2.33)	77.18 (8.27)
12.	Footwear	11.75 (1.62)	10.30 (1.08)	15.46 (1.80)	19.18 (2.06)	24.35 (2.25)	33.63 (2.64)	18.39 (1.97)
13.	Other durable goods	1.91 (0.26)	3.43 (0.36)	6.97 (0.81)	8.28 (0.89)	12.00 (1.11)	4.40 (0.35)	7.31 (0.78)
14.	Rent and taxes	1.69 (0.23)	1.47 (0.15)	1.69 (0.20)	2.49 (0.27)	2.01 (0.19)	0.82 (0.06)	1.85 (0.20)
15.	Miscellaneous goods	49.23 (6.78)	58.85 (6.14)	62.33 (7.24)	69.81 (7.49)	81.73 (7.56)	124.00 (9.75)	70.83 (7.60)
16.	Total expenditure	726.25 (100.00)	958.06 (100.00)	860.53 (100.00)	932.37 (100.00)	1080.39 (100.00)	1271.54 (100.00)	933.22 (100.00)
17.	Per capita monthly expenditure (74-75 prices)	60.52	79.84	71.71	77.70	90.03	105.96	77.77
18.	Per capita monthly income (74-75 prices)	50.45	55.69	73.89	109.97	161.82	229.06	107.46
19.	Per capita monthly expenditure (60-61 prices)	19.90	26.26	23.58	25.55	29.61	34.85	25.50
20.	Per capita monthly income (60-61 prices)	16.59	18.32	24.30	36.17	53.22	75.33	35.34

Note : Figures in parentheses are percentages of total consumption expenditure.

expenditure on costly drinks (soft and well as hard). On the other hand, farm families even at moderate levels of income spend quite a lot on hard drinks. In fact, a big part of rural Punjab has become an addict of alcohol. For relatively costly non-food items like clothing, footwear and miscellaneous goods and services, the per capita expenditure tends to increase with increase in farm size.

On the basis of the foregoing analysis, it becomes fairly evident that the consumption patterns of marginal and small farmers of the Punjab do not reveal a situation of destitution or misery. For essential food items, e.g. cereals, salt and spices, pulses, vegetables and fruit, they are not much behind higher categories of farm families. This shows that even the lowliest of the cultivating households in the rural Punjab are in a position to arrange a square meal. Also the nutritional standards of their food basket are not much inferior. The difference sets in only for superior foods and other "luxury" items. In the matter of per capita expenditure on clothing, footwear, and, miscellaneous goods and services (comprising expenses on health services, education and transport), the lower strata of farm families have a decisive disadvantage. To the extent that some part of this expenditure promotes human capital formation and an assurance for better future living, the picture regarding the future living standards of the lower strata of farm families becomes quite disconcerting. It is only through a policy of active state intervention that such families can be saved from further erosion of their living standards. This can be done in two different ways—first by creating additional income and employment opportunities for such farm families and second, by making cheap education and health services available to them.

Relative Consumption Priorities

The relative consumption priorities are adequately revealed in terms of percentage of total expenditure incurred on each commodity (Table 20.8, figure in parentheses). To obtain a comparative picture, we assign ranks to these percentage figures and compute rank correlation coefficients, between the marginal farmers and the five higher categories (r_{1i} = 2, 3, 4, 5, 6) and also between the overall (average) position and each of the six groups (r_{0i} = 1, 2, 3, 4, 5, 6). These correlation coefficients generally range between 0.95 and 0.99 and none of these is significantly different from unity. This enables us to conclude that all the categories of farm families observe practically the same set of priorities for different items of consumption. Among the food items, cereals, and milk and its products are the items of first importance to all categories of households. Among the non-food items, clothing, fuel and light, and footwear command priorities in that order, by almost all categories of households. In brief, the prosperity brought about by the green revolution in the Punjab does not seem to have disturbed relative consumption priorities of cultivating households so that the consumption behaviour of big farmers is qualitatively not much different from that of the smaller ones. For all practical purposes, they merely consume greater amounts of much of the same bundle of goods and services.

Since the consumption of a large proportion of food items takes place out of home grown stocks, it may be useful to examine the ratio of the two components of consumption (cash and home produced) operating among different categories of farm families. The exercise may serve two objectives—first, to see whether very small and marginal households are in a position to raise output of cereals sufficient for their annual family consumption and second, whether the cash component of expenditure among upper strata of cultivating households is significantly higher compared with households at the lower end. Table 6.9 gives some information on these aspects.

The main constituents of home-produced consumption are cereals, milk and its products and fuel and light. For all categories of households taken together, cash expenditure as a proportion of total expenditure on cereals is about 7.0 per cent; on milk and milk products 3.5 per cent and on fuel and light about 18.0 per cent. These proportions, however, differs widely among different categories of households. For example, the percentage of cash expenditure on cereals is as high as 22.59 for marginal farmers and as low as 1.68 for the biggest farmers; for milk and milk products, the respective figures are 8.09 and 2.48. This clearly shows that the marginal farmers have to purchase nearly one-fourth of their cereal requirements in cash. One of the consequences is that these households are exposed more to the usual price fluctuations in the market compared with other categories of households. Because of non-availability of detailed data on time of purchase and sale of cereals and prices paid and received, it is not possible to work out the full implications of this situation. However, one thing is quite certain. Small farmers deliberately adopt a cropping pattern which tends to maximise their annual net earnings from their limited land areas instead of devoting their entire land to cereal production, as is typical in some traditional agricultural situations. In fact, the cropping patterns of

TABLE 6.9 : Cash Expenditure as percentage of Total Expenditure, Punjab State

Sl. No.	Commodity	Farm Size Group I	II	III	IV	V	VI	All Farms
1.	Cereals	22.59	12.46	5.91	4.66	3.13	1.68	6.76
2.	Pulses	91.23	89.86	80.42	75.76	71.40	76.94	79.80
3.	Milk and its products	8.09	6.29	4.73	2.09	2.01	2.48	3.50
4.	Edible oils	83.05	84.31	76.25	80.12	72.41	86.76	78.83
5.	Meat, egg, fish, etc.	88.51	82.42	85.44	82.63	80.08	89.15	83.67
6.	Vegeatables, fruits, etc.	77.64	84.42	77.09	76.39	78.38	82.69	78.83
7.	Sugar, gur, etc.	82.41	74.07	65.76	68.70	70.88	75.47	70.66
8.	Salt and spices	100.00	100.00	100.00	100.00	100.00	100.00	100.00
9.	Beverages	20.46	16.13	15.93	19.29	17.43	21.25	17.81
10.	Fuel and light etc.	100.00	100.00	100.00	100.00	100.00	100.00	100.00
11.	Clothings	100.00	100.00	100.00	100.00	100.00	100.00	100.00
12.	Footwear	100.00	100.00	100.00	100.00	100.00	100.00	100.00
13.	Miscellaneous goods	100.00	100.00	100.00	100.00	100.00	100.00	100.00
14.	Durable goods	100.00	100.00	100.00	100.00	100.00	100.00	100.00
15.	Rent and taxes	100.00	100.00	100.00	100.00	100.00	100.00	100.00
16.	Total expenditure	45.79	41.46	39.70	38.87	40.09	47.40	40.80

marginal and small farmers are as much commercialised and market oriented as of any other category. The only disadvantage that they have to suffer is because of the invariable rise in the price of cereals at the time that they have to purchase them. The extension of public distribution system for the lower strata of the rural poor is the only way that these sections can be protected against this disadvantage.

On commodities such as edible oils, meat, egg and fish, vegetables and fruit, cash expenditure makes a very high percentage of total expenditure. For clothing, footwear, miscellaneous goods and services, and durable goods among the non-food items, the entire expenditure is in cash for all categories of households.

In general, about 40 per cent of total consumption expenditure is made in cash, considering all households together. In other words, three-fifths of the total consumption of cultivating households in the Punjab is made possible by farm itself. The proportion of home produced consumption is much lower for the marginal and small cultivating households.

Distribution of Consumption Expenditure

For analysing the distribution of consumption expenditure among the various strata of cultivating households, the households are rearranged into decile groups, in ascending order of their net operated area (rather than in ascending order of their total household expenditure) and the percentage share of each decile in total as well as in individual commodity expenditure is worked out. To gain greater insight into the distribution of consumption expenditure at extreme ends, the first and the last deciles have been further divided into four equal sub-groups. The above arrangement is preferable since our basic interest lies in building the production-income-consumption nexus of cultivating households by their holding size and not of expenditure classes as such.

Table 6.10 gives commoditywise distribution of consumption expenditure. A number of interesting facts emerge from this table. Firstly, all decile groups, including those operating at the very bottom, do have a share in the consumption of most of the commodities studied. In particular, even in the consumption of costly items such as meat, egg and fish, clothing, footwear, etc, the bottom 2.5 or 5.0 per cent of households report some expenditure *albeit* very small. Secondly, the distribution of expenditure on basic necessities such as cereals, pulses, edible oils, salt and spices, etc, conforms almost to a perfectly egalitarian norm. For these items, the Gini's Concentration Ratios are extremely small. It is fairly clear that in the consumption of basic food items, the lower strata of the cultivating households in the Punjab claim a share that is more or less proportionate to their population size. Thirdly, expenditure on meat, egg and fish, milk and its products and, to some extent, beverages among the food items, and durable goods, clothing, footwear and miscellaneous goods and services (which includes education, health, etc) among the non-food items, is distributed more in favour of the upper deciles. As a matter of fact, the tilt sets in right at the bottom so that between any two consecutive deciles (or two sub-groups within a decile), the share of the succeeding decile is almost invariably greater. It is, however, equally important to bear in mind that for these commodities, the bottom 50 per cent of the consuming population have less than their proportionate share. On the other hand, the share of the top decile in general, and among them of the top 2.5 per cent, is proportionately much larger than their numerical strength among consuming units. The Gini's Concentration Ratios for these important items are relatively much higher than those for basic food

TABLE 6.10 : Distribution of Consumption Expenditure by Decile Groups, Punjab State

	Percent Share of Consumption Expenditure							
Decile Group	*Percentage of Consumers*	*Cereals*	*Pulses*	*Milk and Milk Products*	*Edible Oils*	*Meat, Eggs and Fish*	*Fruit and Vegetables*	*Sugar and Gur*
0-2.50	1.90	1.64	1.40	1.26	2.17	0.75	1.65	1.54
2.50-5.00	2.15	1.84	1.79	1.21	1.92	1.05	1.91	1.85
5.00-7.50	1.84	1.82	1.56	1.68	1.85	0.64	1.57	1.55
7.50-10.00	2.18	2.08	2.15	1.30	2.62	1.14	1.79	1.74
10-20	8.08	8.00	7.44	6.89	9.51	5.92	7.43	7.62
20-30	8.73	8.65	8.11	7.04	8.17	5.22	7.26	8.11
30-40	9.19	8.77	8.73	7.50	9.35	4.18	8.10	8.55
40-50	9.35	9.27	8.12	8.87	10.58	7.57	8.70	9.04
50-60	10.07	9.99	9.31	10.05	8.26	6.83	9.47	9.90
60-70	10.69	10.30	10.17	10.22	10.42	10.33	10.43	10.04
70-80	11.39	11.68	12.07	13.33	9.68	9.47	11.53	11.73
80-90	11.49	11.66	12.17	13.85	11.96	15.00	12.30	12.71
90-92.50	2.71	2.93	4.27	3.25	3.70	4.74	3.91	3.27
92.50-95.00	3.05	3.46	4.28	3.65	3.37	5.07	3.68	3.45
95.00-97.50	3.47	3.59	3.92	4.32	2.38	8.53	4.23	4.18
97.50-100.00	3.72	4.15	4.52	5.00	4.04	13.49	6.06	4.73
Total	100.00	100.00	100.00	100.00	100.00	100.00	100.00	100.00
Gini's Ratio		0.0220	0.0394	0.0996	0.0623	0.2893	0.0775	0.0549

TABLE 6.10 (*Contd.*)

Decile Group	*Percent Share of Consumption Expenditure*							Total Household Expenditure
	Salt and Spices	Beverages	Fuel and Light	Clothing	Footwear	Misc. Goods	Durable Goods	
0-2.50	2.15	1.27	1.64	0.77	1.05	1.40	0.29	1.38
2.50-5.00	1.89	1.73	1.71	1.19	0.52	1.40	0.21	1.50
5.00-7.50	1.95	1.33	1.83	0.96	1.02	1.39	3.21	1.59
7.50-10.00	2.17	1.55	2.05	1.75	2.18	1.34	0.12	1.88
10-20	7.99	7.24	8.34	4.93	5.05	7.19	0.18	7.19
20-30	8.52	7.54	8.07	5.18	5.09	6.95	5.52	7.39
30-40	9.00	8.50	9.04	6.00	6.77	7.75	15.25	7.95
40-50	9.79	8.72	9.25	8.25	8.12	8.92	3.24	8.99
50-60	9.65	9.46	10.39	11.29	8.21	8.57	1.66	9.86
60-70	10.48	10.05	10.33	10.22	11.54	11.06	10.92	10.35
70-80	11.08	11.18	11.69	11.70	13.97	11.26	14.34	12.20
80-90	11.79	11.95	11.14	13.95	15.68	15.22	3.80	12.97
90-92.50	2.59	2.88	2.85	5.12	4.47	3.90	17.52	3.44
92.50-95.00	3.28	3.18	2.84	5.18	4.19	3.25	13.23	3.69
95.00-97.50	3.61	4.74	3.94	5.13	4.62	3.06	8.32	4.25
97.50-100.00	4.04	8.66	4.88	8.37	7.47	7.33	2.19	3.49
Total	100.00	100.00	100.00	100.00	99.98.00	100.00	100.00	100.00
Gini's Ratio	0.0595	0.1004	0.0240	0.2034	0.1946	0.1414	0.3001	0.0834

items. Lastly, the distribution of total household expenditure is much less skewed than that of total household income.[8] This is so because:

(a) the consuming units with limited land base and total household income are spending (or are obliged to spend) more than their current earnings, and are thus deficit units; and
(b) the consuming units with a big land base are consuming less than their current earnings and are thus surplus units.

To some extent, the deficit of the bottom two deciles or so might be explained in terms of overestimation of their consumption expenditure or underestimation of their annual income. There are some items of consumption, earned informally through wage employment or crops or vegetables raised on own farm, which are duly included in consumption account but are generally excluded from income account. Obviously this tendency operates much more with operators of limited land base. The real deficit of the bottom two deciles or so is, therefore, much smaller than what appears in our estimates. However, our data show that these bottom two deciles record a deficit in all the districts of Punjab, irrespective of the general level of their agricultural development.

Negative savings for a year or two may be a normal feature with marginal and small farmers. However, if the deficit occurs year after year, it can lead to many consequences. One such consequence could be that their debt obligations pile up over time and they are obliged to mortgage or sell their land and join the ranks of rural and/or urban proletariat, or else become tenant cultivators. Alternatively, while holding on to land, these households may be obliged to mortgage their jewellery or other assets.

Unfortunately, however, no systematic data are available on these important aspects and only impressionistic views go around. It is, for example, generally believed that the Punjab farmer is madly in love with his land and therefore the possibility of his selling or mortgaging land to clear consumption debt is extremely remote. Consequently many knowledgeable people believe that there are very rare cases of land sales taking place under duress.

The obvious consequences of large scale land sales by marginal and small farmers would be an increase in the number and proportion of landless labour and decline in their share of total owned area. The data given both by the Population Census and the All-India Rural Labour

Enquiry indicate that the proportion of (male) agricultural labour recorded a significant increase during the sixties and the early seventies. Thus, according to the Population Census, the percentage share of male agricultural labourers in total workforce increased from 9.65 per cent to 20.02 per cent between 1961 and 1971, and from 20.02 to 22.61 during 1971 and 1981. According to the Rural Labour Enquiry, labour households in the Punjab (including Haryana) increased by 33 per cent during 1964-65 to 1974-75. Similarly, the NSS data also show that during 1961-71 the proportion of cultivators operating less than 5 acres has perceptibly declined, although their share of area operated has slightly increased. Unfortunately, however, it is not possible to know if the increase was due to pauperisation of small and marginal farmers, influx of workers from other sectors or immigration of labour from outside the state. The Census data indicate that the proportion of cultivators has declined only marginally and that most of the increase is because of a section of workers from the services sector going into the agricultural workforce. Once again, the data on assets also do not lead to any conclusive evidence about pauperisation. For example, the RBI data on assets actually show that the asset position (including land) of the first two deciles has actually improved although their share in the total assets has significantly declined between 1961 and 1971.

To sum up, it is very difficult to draw any definite conclusions from the available data to support or reject the distress land sale and the pauperisation hypotheses.

However, whatsoever the magnitude and nature of deficit, it should be a matter of great concern that even in the heart of the green revolution, the marginal and small farmers find it quite difficult to make both ends meet. The implication is that as long as the marginal and small farmers of Punjab operate with their limited land base, as at present, and if their income-consumption deficit persists over time, there is a real danger that the process of losing land may set in the near future, if it has not started already. Positive policy measures will have to be taken to supplement the income of marginal and small farmers from non-farm productive activities, if these people are not to be driven to swell the ranks of rural landless labour.

Expenditure Elasticities

In this section, an attempt is made to estimate expenditure elasticities for each of the commodity groups mentioned earlier. For this purpose, bivariate Engel Functions were fitted with per capita commodity expenditure (y) as the dependent variable and per capita total expenditure (x) as the explanatory variables. It is well known that such per capita formulations imply neutrality as regards the economies of scale in consumption.

For estimating Engel Functions, some procedural adjustment was necessary. In the first instance, the wide intra-group variations in total as well as commodity expenditures had to be reckoned with Following Klein (1969),[9] the cultivating households were first divided into as many as fifteen categories (against only six considered so far), on the basis of net area operated by them, and then, for each such household group, the mean level of per capita total as well as commodity expenditures were computed. This gave us fifteen pairs of observations on *x* and *y*. Our next step was to run weighted regressions between *x* and *y*, the number of households behind each pair of *x* and *y* being used as the relevant weight. The main advantage of the above procedure is that the household classification based on operated area is intertwined with that based on total consumption expenditure, and consequently, inferences derived from the analysis of the latter would have, *mutatis mutandis,* their counterpart in the former.[10]

After a careful scrutiny of the available mathematical formulations, we settle on the following four forms:

Semi-Log . .

$$y = a + b \log_e x \qquad \ldots (1)$$

Hyperbolic . . .

$$y = a - b \frac{1}{x} \qquad \ldots (2)$$

Log-Linear . . .

$$\text{Log}_e y = a + b \,\text{Log}_e x \qquad \ldots (3)$$

Log-Inverse

$$\text{Log}_e y = a - b \frac{1}{x} \qquad \ldots (4)$$

The functions selected by us satisfy most of the economic and statistical conditions usually associated with Engel functions. The estimated Engel Functions are given in Table 6.11. Corresponding to each functional type the expenditure elasticities (e's) and the co-efficients of determination (r^2) are also given in the Table. Strictly speaking, the r^2s obtained from various forms are not comparable unless the dependent variable (y) is suitably transformed, say, on the lines suggested by Box and Cox.[11] However, in actual practice, such an adjustment makes a very negligible effect on the magnitude of r^2.[12] Under these circumstances, it is unfair to accept or reject a particular form unless the differences in r^2s are much too wide. In our case, r^2 differs within very narrow margins, from one functional form to the other, for almost all commodity-groups. We have, therefore, decided to retain all the four forms for each commodity group.

TABLE 6..11 : Engel Functions for Cultivating Households, Punjab State

	Commodity Group	*Function Type*	*Estimated Value of*				
			Intercept (a)	*Slope (b)*	*t (b)*	r^2	*e*
	1	2	3	4	5	6	7
1	Cereals and cereal products	Semi–Log	–117.27	45.78	2.90*	0.18	0.236
		Hyperbolic	239.40	–40231.00	2.79*	0.16	0.227
		Log–Linear	3.70	0.23	2.89*	0.17	0.231
		Log–Inverse	5.49	–204.06	2.80*	0.16	0.223
2	Pulses	Semi–Log	–76.11	15.12	298*	0.18	0.564
		Hyperbolic	41.01	–12665.27	2.70*	0.15	0.519
		Log–Linear	–0.31	0.53	2.78*	0.16	0.530
		Log–Inverse	3.76	–441.72	2.53*	0.14	0.485
3.	Milk and milk products	Semi–Log	–2385.57	–394.33	14.75*	0.85	1.320
		Hyperbolic	705.73	–363486.88	15.61*	0.86	1.337
		Log–Linear	–3.66	1.37	12.45*	0.80	1.370
		Log–Inverse	7.10	–1270.42	13.09*	0.83	1.405
4.	Edible oils	Semi–Log	53.35	4.90	0.77	0.01	0.245
		Hyperbolic	15.36	4133.12	0.71	0.01	0.227
		Log–Linear	5.01	–0.30	0.92	0.02	0.300
		Log–Inverse	2.64	265.13	0.88	0.02	0.291
5.	Meat, egg and fish	Semi–Log	–179.55	27.60	6.13*	0.49	3.314
		Hyperbolic	35.17	–23967.98	5.60*	0.44	3.165
		Log–Linear	–18.85	3.05	5.75*	0.46	3.050
		Log–Inverse	4.96	–2747.40	5.65*	0.44	3.020
6.	Vegetables and fruit	Semi–Log	105.37	20.41	3.15*	0.20	0.608
		Hyperbolic	52.97	–17322.70	2.90*	0.17	0.568
		Log–Linear	–0.32	0.56	2.87*	0.17	0.560
		Log–Inverse	4.02	–477.10	2.66*	0.15	0.524
7.	Sugar and gur	Semi–Log	–278.57	51.17	8.55*	0.65	0.733
		Hyperbolic	121.58	–46275.38	8.34*	0.63	0.730
		Log–Linear	–0.69	0.72	8.37*	0.64	0.720
		Log–Inverse	4.97	–658.74	8.30*	0.63	0.724

(Contd.)

1	2	3	4	5	6	7
8. Salt and spices	Semi–Log	2.56	2.40	0.54	0.01	0.127
	Hyperbolic	21.76	–2552.86	0.63	0.01	0.149
	Log–Linear	2.01	0.13	0.58	0.01	0.130
	Log–Inverse	3.07	–139.05	0.67	0.01	0.154
9. Beverages	Semi–Log	–432.17	69.78	7.49*	0.59	1.630
	Hyperbolic	110.74	-6.01×10^4	6.72*	0.53	1.544
	Log–Linear	–5.84	1.40	9.18*	0.68	1.400
	Log–Inverse	5.11	-1.25×10^4	8.49*	0.64	1.374
10. Fuel and light	Semi–Log	–33.74	11.82	2.59	0.14	0.253
	Hyperbolic	58.25	-1.03×10^4	2.46	0.13	0.243
	Log–Linear	2.11	0.25	2.63	0.15	0.250
	Log–Inverse	4.09	–224.24	2.53	0.14	0.246
11. Clothing	Semi–Log	–1203.83	191.50	9.60*	0.70	1.921
	Hyperbolic	288.72	-1.69×10^5	8.62*	0.65	1.865
	Log–Linear	–7.62	1.79	9.46*	0.69	1.790
	Log–Inverse	6.34	–1604.65	8.99*	0.67	1.763
12. Footwear	Semi–Log	–238.36	38.63	5.42*	0.43	1.571
	Hyperbolic	62.97	-3.43×10^4	5.18*	0.40	1.536
	Log–Linear	–7.58	1.57	5.17*	0.40	1.570
	Log–Inverse	4.72	–1421.72	5.10*	0.39	1.562
13. Miscellaneous goods	Semi–Log	–933.07	154.52	7.87*	0.61	1.300
	Hyperbolic	291.97	-1.37×10^4	7.31*	0.57	1.267
	Log–Linear	–3.76	1.25	7.19*	0.57	1.250
	Log–Inverse	5.99	-1.12×10^3	6.87*	0.54	1.230
14. Durable goods	Semi–Log	–135.69	23.90	1.78	0.07	0.885
	Hyperbolic	54.40	-2.45×10^4	2.03	0.09	1.000
	Log–Linear	–2.10	0.77	1.43	0.05	0.770
	Log–Inverse	4.08	–802.21	1.71	0.07	0.914

Note : *e* denotes Expenditure Elasticity and * implies that the slope (b) is significant at 0.01 level.

A number of interesting facts emerge from the estimated functions. Firstly, the statistical significance of the slope (b) does not depend on the form chosen. If significant, it is so for each of the four forms; if non-significant, again it is so for each of the four forms.[13] Secondly, while for each commodity group, r^2 varies only marginally from one functional form to the other, it varies widely from one commodity-group to the other. For example, r^2 ranges between 0.80 and 0.86 for milk and milk products and is as low as 0.01 for edible oils. It is fairly high for clothing, footwear, meat, egg and fish, beverages, miscellaneous goods and services, etc. Further, r^2 is fairly low for cereals, salt and spices, pulses, etc. These results have interesting implications. The level of total per capita expenditure on milk and milk products, clothing, footwear, beverages, meat, egg and fish, miscellaneous goods and services, etc. is more intimately tagged with the per capita total expenditure of the household.[14] On the other hand, the per capita expenditure on essential food items (e.g., cereals and cereal products, pulses, salt and spices) is much less influenced by per capita total expenditure. It can be explained perhaps by the fact that the per capita consumption of basic food items is already fairly high,[15] so that the rising level of total per capita expenditure seeks outlets relatively more in the consumption of non-food items such as clothing, footwear, beverages, etc. The general conclusion is obvious enough that the consumption standards of cultivating households are under-going qualitative improvements. Thirdly, for each commodity-group, the value of expenditure elasticity (e) does not vary much from one functional form to the other. As a matter of fact, for essential food items (e.g., cereals and cereal products, pulses, salt and spices, sugar and gur, etc), the elasticities are almost invariant to functional form. Incidentally, this lends further support to our choice of functional forms. Fourthly, in the matter of commodity priorities as revealed by the value of expenditure elasticity, meat, fish and egg have the highest value of expenditure elasticity excessding 3.0, clothing has the next highest value of nearly 1.8 footwear, beverages and durable goods share among themselves the next three positions with expenditure elasticity figures exceeding 1.5 in most cases. As expected, the lowest expenditure elasticity around 0.22 is shown by cereals and cereal products. This clearly shows that the general consumption patterns of the cultivating households of Punjab are strikingly different from those that are usually associated with poor, subsistence economies. On the one hand, the cultivating households of the Punjab do not suffer from cereal deficiency; hence expenditure elasticity for cereals is the lowest. On the other hand, there is the tendency to spend additionally on meat, egg and fish an amount which is much more than the percentage addition to income. Also, clothing and footwear command fairly high priorities in spending incremental incomes.

Households Below Poverty Line

In this section, a brief attempt is made to work out the percentage of cultivating households living below the poverty line. To avoid procedural complexities, simple head count method has been used to estimate the incidence of poverty. Following Bardhan[16] and the Seventh Finance

Commission Report,[17] households with less than Rs. 15.90 as per capita monthly expenditure at 1960-61 prices have been designated as 'poor households'. The corresponding figure for 1974-75 works out to be Rs. 48.33.[18] For well known reasons, working with a single year data, per capita expenditure, as the basis for demarcating poverty line, is preferable to per capita income. The poverty figures are given in Table 6.12.

TABLE 6.12 : Cultivating Households below Poverty Line (per capita Monthly Expenditure is 48.33 at 1974-75 Prices = Rs. 15.90 at 1960-61 prices)

Farm Size Category	*No. of Sample Households*	*No. of Poor Households*	*Average Per Capita Monthly Expenditre of Household below Poverty Line*	*(Rs.)*
I	140	44 (31.43)	Bottom 20 per cent	30.00
II	326	77 (23.62)	Next 20 per cent	36.40
III	404	77 (19.06)	Next 20 per cent	40.35
IV	409	47 (11.49)	Next 20 per cent	43.60
V	310	25 (8.06)	Next 20 per cent	47.01
VI	74	4 (5.41)	All Poor households	39.52
All Farms	1663	274 (16.48)		

Note : Figures in parentheses are poor cultivating households as a percentage of all cultivating households.

Looking across farm size categories, it is striking to note that about one-third of the marginal farmers (tilling less than 2.5 acres of land) are living below the poverty line. It is ironic that despite a tremendous advance in technology, many of the marginal farmers in Punjab are still unable to take out a minimum living. It is equally disturbing that about 24 per cent of small farmers (tilling between 2.5 to 5.0 acres of land) are also living below the poverty line.

Table 6.12 shows that poor households belong to all farm size categories although their percentage declines sharply with increase in land holding. Quite a few 'poor' small and marginal farmers find it difficult to earn adequate income from their limited resource base. In their case, low income is the main cause of their low spending. The observed poverty in the case of higher categories is due largely to a few abnormal observations that usually arise in any large scale study on consumption. These abnormalities are obvious in the present case since the incomes reported by the so-called 'poor' in higher categories are a multiple of their reported consumption.

V. CONCLUSIONS AND POLICY MEASURES

The advent of the green revolution in Punjab has brought overall prosperity to its peasantry. Thanks to the creation of an assured irrigation base and its fairly equitable distribution, all categories of cultivators have been able to record substantial increase in their output and income through the adoption of new technology. However, the gains of the new technology have been distributed more or less in proportion to the initial land holding position. Since the land holdings are distributed in a very skewed manner, the distribution of gains has also been quite inequitable. It goes to the credit of marginal and small farmers of Punjab that despite their limited land base, they have been able to record almost as much total crop output and farm business income per acre as their bigger counterparts. Thus they are able to achieve primarily by achieving a much higher cropping intensity through a rational year-round utilisation of their family labour.

A higher cropping intensity compensates them for slightly lower yield rates for individual crops such as rice which are currently at the threshold of a new technological breakthrough. It is, however, notable that in the central Punjab plains, the traditional inverse relationship between farm size and productivity stands reversed. In this region, in spite of higher cropping intensity, the marginal and small farmers are unable to obtain higher output per acre. Their production efficiency notwithstanding, these farmers are unable to earn adequate per capita income from crop production because of their small land base. They are, therefore, obliged to engage themselves in a wide range of non-farm activities and are able to obtain higher per capita earnings from such activities. Consequently, the distribution of total household income (of which income from non-farm activities is an important constituent for marginal and small farmers) is much less inequitous than land or farm income distribution. Unfortunately, however, in spite of putting in very hard labour both in farm and non-farm activities quite a few amongst them are not able to save themselves from the clutches of poverty. It is indeed a disquieting feature of the Indian agrarian situation that even in the heartland of green revolution about one-third of the marginal farmers and about one fourth of the small farmers are living below the poverty line.

Some suggestions can be given for improving the economic condition of marginal and small farmers.

All efforts should be made to further strengthen their crop production base. To enable them to participate fully in future technological break-throughs, improved inputs should be provided to them through the creation of exclusive co-operatives. These cooperatives should also provide pooled services ultimately leading to the emergence of full-fledged co-operative farming. There is perhaps no other alternative for reducing the severe scale diseconomies which afflict them as individual production units. It must be stressed that the pre-condition for the success of these co-operatives is good and honest management, and the assurance that the subsidised inputs made available to these co-operatives do not leak into the coffers of the bigger peasants.

Simultaneously, efforts should also be made to involve marginal and small farmers in allied agricultural and non-agricultural activities on an increasing scale to provide them with higher productive employment. This will involve creation of a big infrastructure for making available the necessary inputs at reasonable prices, collection of produce, its processing and marketing.

Notwithstanding the potential gains from the new agricultural technology and allied activities mentioned above, it needs to be underlined that it is not possible to make a decent living in agriculture without an adequate land base. True the green revolution has made a dent on poverty of small farmers by reducing the economically viable size of holding. But there are obvious limits to this process as diseconomies of small scale production tend to become extremely severe. Hence, ultimately methods have to be found to increase the land base of marginal and small farmers. This can only be done through radical land redistribution. Unfortunately, the economic and political power of the rich peasantry is so great and the peasant movements are so weak that this does not seem to be on the agenda.

In the ultimate analysis it is quite erroneous to believe that all problems of rural poverty can be solved within agriculture. One of the main reasons for rural poverty is overpopulation in agriculture combined with inequality in land distribution. It is, therefore, essential to withdraw labour force from agriculture into non-agricultural occupations and industry. Ultimately, the real solution of the rural and urban poverty in developing countries lies in the rapid industrialization and diversification of the economy. This in turn requires capital formation at a rapid rate. However, mobilisation of surplus for capital formation from rural areas meets with fierce opposition from the well entrenched rural interests. The type of polity we have seems incapable of breaking their stranglehold.

The Green Revolution in Punjab, as elsewhere in the country, has provided a breathing space by productively absorbing increasing labour force into agriculture. But the dynamics of the situation is such that all increases in rural labour force cannot be absorbed into agriculture for all times to come. The vision of the policy makers will primarily depend on whether or not they can bring about the needed economic transformation along with taking short term measures for radical redistribution of assets and institution of new organisational innovations.

Note : Reproduced from *Economic and Political Weekly*, Vol. XVII, No. 20 & 21, May 15 and 22, 1982, pp. 1613.

NOTES

1. (a) Bardhan, P.K. 1973, "The Incidence of Rural Poverty in the Sixties", *Economic and Political Weekly,* Annual Number, February.
   (b) Byres, T. J. 1972. "The Dialectic of Indian Green Revolution", *South Asian Review,* Volume 5, No. 2.
   (c) Griffin, K. and Khan, A., 1976, "Poverty an Landlessness in Rural India, ILO (Mimeo.).
2. Krishna Bharadwaj, *Production Conditions in Indian Agriculture*, Cambridge University Press, 1974.
3. The percentage of net sown area irrigated for the six farm size categories is: 80.76, 82.49, 83.03, 83.46, 83.34, 87.33, respectively.
4. It was only 2.44 lakh nutrient tonnes in 1974-75 against 3.07 lakh nutrient tonnes in 1973-74.
5. As a matter of fact for some districts, e.g. Bhatinda and Sangrur, the farm sizewise differences are almost non-existant.
6. Table 6.4 gives the impression that the yield rates for cotton also increase uniformly as we move upwards along the farm size ladder. This is not quite so. The yield rate for marginal farmers alone is lower while for all other categories, yield rates are almost invariant.
7. For many conceptual difficulties inherent in a single year data, it is extremely difficult to evaluate the uptodate value of the stock of capital with a cultivating house-hold and hence depreciation cannot be worked out without making questionable assumptions.
8. Gini's Concentration Ratios for household income and household expenditure are 0.336 and 0.084 respectively.
9. L.R. Klein, *An Introduction to Econometrics*, Prentice Hall, 1969, pp. 56-57.

10. Our estimated functions based on varying levels of per capita total expenditure are not quite the same as those based on varying levels of per capita household income. There is no one-to-one correspondence between per capita income and per capita expenditure, especially because towards the lower end of the income ladder, many households spend beyond their current earnings. Apart from the above, in analysing the consumption behaviour of cultivating households, there is great intutive appeal to use per capita expenditure as the explanatory variable in preference to per capita income since the consumption standard of house holds has the closest resemblance with per capita expenditure rather than per capita income earned during a single year.
11. G.E.P. Box and D.R. Cox, "An Analysis of Transformation", *Journal of the Royal Statistical Society,* Series B. 1964, pp. 211-48.
12. D.B. Gupta, *Consumption Patterns in India,* Tata McGraw Hill, 1973, Ch. 4.
13. The exceptions are negligible.
14. The result on milk and milk products needs some clarification. Firstly, this single commodity group makes a sizable part of the total consumption expenditure. Secondly, the consumption of milk and milk products is mostly out of home produced stock so that negligible cash expenditure is involved. In view of the above, very high value of r^2 points to the tendency of higher retentions for home consumption out of the increasing quantity of milk production, the latter having important bearing on household income and expenditure.
15. This is confirmed separately by the fact that per capita expenditure on cereals and cereal projects, pulses, salt and spices. etc. shows negligible variation as we move up on the ladder of expenditure classes. On the contrary, such variations are fairly wide in the case of clothing, footwear, beverages, etc.
16. Bardhan, P.K., "On the Incidence of Poverty in Rural India in the Sixties", in *Poverty and Income Distribution,* T.N. Srinivasan and P.K. Bardhan (*eds*), 1974.
17. Government of India, *Report of the Seventh Finance Commission,* 1978.
18. Consumer Price Index for Agricultural Labourers in Punjab (including Delhi, Haryana and Himachal Pradesh), prepared by Labour Bureau, Government of India, has been used for the purpose.

7

GROWTH IN THE CONTEXT OF UNDERDEVELOPMENT

Case of Dryland Agriculture

V.M. RAO

I. INTRODUCTION

A substantial breakthrough in dryland agriculture is a key component in the agricultural strategy proposed for the Eighth Plan (Planning Commission, 1990). This is important not only for sustaining agricultural growth over the coming decades but, even more, to locate agricultural growth processes in vast areas which have remained largely stagnant so far. Nurturing growth processes in such areas is a delicate and time taking development task. While the generation of new technologies for dryland agriculture is a formidable enough challenge to be faced, the effective adoption of these technologies by the masses of peasants to support a steady and viable growth processes is immeasurably more so. Changes of such an order are in the nature of thoroughgoing transformation in rural societies covering individual attitudes and behaviour, community structures and the network of hierarchies and relationships influencing their economic performance.

The purpose of this paper is to provide glimpses of the barriers to growth in the dryland communities. Its focus is on the peasants in these communities and not on the important—but, for the theme of this paper, somewhat peripheral—technical issues relating to improvements in dryland agriculture. More specifically, we present findings indicative of the depressive influence exerted by the environment—physical and socio-economic—on the peasants' participation in the growth processes. The term 'underdevelopment' used in the title is merely a convenient and expressive way of describing this feature of the environment; there is no intention to deal with the theoretical issues and debates raised in the large and growing literature on underdevelopment though our findings could be of more than cursory interest to social science researchers trying to comprehend the phenomenon of under development.

The environment in dryland agriculture areas has three major growth-depressing characteristics—harsh physical conditions, low priority in development policies and investments and inability to complete with the stronger and more modernised and dominant parts of the economy. It is useful before proceeding further to have a brief look at these characteristics.

(a) Regarding the physical conditions, the following is an assessment by the Planning Commission of the conditions in the 'Southern Plateau and Hills Region'—the region containing the dryland areas in Karnataka from which are drawn the findings reported in this paper: "a large, typically semi-arid zone having 81 per cent of dryland farming with about 50 per cent cultivated area and low cropping intensity of 111 per cent. Low value cereals and minor millets predominate due to rainfed farming. Low level of irrigation has led to unstable and low productivity" (Planning Commission, 1989). According to an another expert study, the dryland areas in this region are "subject to physical constraints and degraded soil and water resources. They are of intermediate phase requiring concentrated intensive efforts for protection/development of land and water" (Basu and Rajagopalan, 1990).

(b) As regards the low priority in development policies and investment, it is reflected well in 'degraded soil and water resources'—an outcome of past neglect and indifference towards the need for policies for 'concentrated intensive efforts for protection/development'. Significantly, the approaches like agro-climatic zonal planning and watershed development—approaches providing the conceptual-cum-analytical support for the making of such policies—appeared on the scene only in the mid-eighties and after. It is tempting to associate their appearance with the growing scepticism of recent years about the capacity of the green revolution areas, by themselves, to sustain agricultural growth in India. Often, the neglect of agriculture in our development strategy is sought to be evidenced by presenting data on the level and tempo of Plan allocations for agriculture. In the case of dryland agriculture, the absence of adequate planning approaches to land and water resources until very recently is an even more eloquent symptom of the Planners' attitude. The Planning Commission document mentioned above has identified six thrust areas for priority policy attention for the 'Southern Plateau and Hills Region' (see Table 7.1). It has also recommended 16 priority programmes classified by the thrust areas. It would seem that seven out of these 16 programmes depend critically for their effectiveness on the existence in the field of an adequate planning approach and that these, in turn, affect the success of other priority programmes and, also, of the agricultural strategy as a whole underlying the identified programmes. Table 7.1, column 2 lists these 'critical' programmes. It is our contention that the present rudimentary state of planning systems for agro-climatic zones and watersheds-owing to the limited policy attention paid to dryland agriculture until almost the mid-eighties—is a telling sign of the weak policy environment of dryland agriculture.

(c) Probably the strongest and most pervasive 'under development' force impinging on dryland economy and on the conditions of communities subsisting in dryland areas is their inability to cope with the stronger parts of the mainstream economy. An excellent illustration of the unequal competition between the two is provided by a recent study of cotton crop in the Rayalaseema area in Andhra Pradesh over the period 1800 to 1980. The aim of this study was to "throw light on the steadily declining trends in the share of Rayalaseema districts as centres of cotton production". The study notes that "since the Vijayanagara times cotton had been one of the chief commercial crops of Rayalaseema.... Colonial rulers encouraged its cultivation by numerous incentives and experimented (with) new varieties of cotton on the black tracts of Rayalaseema during the 19th century. However, area under cotton declined steadily in the present century. The area, productivity and prices fluctuated frequently from year to year in the post-independence period and the importance of Rayalaseema as a cotton producing region declined sharply after 1965 with the advent of the high-yielding and irrigated varieties of cotton in the other regions of Andhra Pradesh and India.... Rayalaseema region enjoyed the advantages of producing mostly dry crops up to the end of 19th century; but could not stand out in the face of commercialisation of agriculture, in the other regions in this century" (M. Atchi Reddy and Y. Sarojini, 1990).

TABLE 7.1 : Thrust Areas and 'Critical' Programmes for Dryland Areas*

Thrust Area		*'Critical' Programmes*		*Other Priority Programmes*
Dryland Farming	(i)	Delineation of macro and micro watersheds for water conservation and integrated development	(i)	Development of early fertiliser-responsive varieties of dryland crops
	(ii)	Land levelling and bunding with agro-forestry and crop production dovetailed with employment programmes		
Minor Irrigation	(iii)	Restoration of tanks by deepening embanking etc.	(ii)	Improved irrigation delivery systesms, on-farm practices and adoption of water saving devices.
	(iv)	A large programme for dug-cum-bore wells		
Crop diversification	(v)	Development of arid/semi-arid horticulture	(iii)	Diversion of area from coarse grains to oil seeds
			(iv)	Cultivation of soyabean/safflower on residual moisture
Delivery Systems		—	(v)	Supply of quality seeds for lead crops
			(vi)	Transit storage godowns for fruits and vegetables at strategic points
Soil Reclamation	(vi)	Master plan for drainage in problem areas	(vii)	Conjunctive use of water in irrigation command areas
Livestock	(vii)	Augmentation of fodder production by integrating it with wasteland development	(viii)	Upgradation of cattle buffalo breeds
			(ix)	Support to poultry farming

Note : * 'Critical Programmes are those which depend on adequate planning systems for agro-climatic zones and watersheds. Other programmes depend, in turn, on 'critical' programmes.

Source : Planning Commission (1989).

It is obvious that pockets of growth now in the process of emerging in this inhospitable 'underdevelopment' environment would be of particular interest to those concerned with the analysis and policy-making for dryland agriculture. A number of studies done in the Institute for Social and Economic Change have shown that Karnataka agriculture, including its dryland part, experienced a fair measure of growth over the period extending from the mid-fifties to the late seventies. The supporting evidences presented in these studies include (a) comparison of aggregate agricultural growth between Karnataka and India, (b) comparison of growth rates in selected dryland crops between the State and India, and (c) district-cum-cropwise analysis to locate pockets of growth within Karnataka. More interestingly, recent studies of Karnataka agriculture have revealed evidence of declaration and stagnation after the late seventies through the scenarios do differ across crops. For example, Bislaiah has presented convincing evidence that the rate of growth in productivity which accelerated in the period 1968-69 to 1978-79 (period E2) compared to the period 1955-56 to 1967-68 (E1) experienced a steep fall in the next period, viz., 1979-80 to 1987-88 (E3) (see Table 7.2). Thus Karnataka's experience would be particularly relevant for this paper since the State has witnessed a phase of agricultural growth followed by a phase of stagnation in the last decade. Hopefully, this experience would provide clues to both what promotes growth and what retards it.

TABLE 7.2 : Change in Productivity of Major Crop Groups and Crops

	Per Cent Change during	
	E2 over E1	*E3 over E2*
Cereals	47	11
Pulses	34	–9
Oil seeds	No change	15
Rice	40	11
Ragi	31	17
Jowar	73	3
Bajra	57	14
Groundnut	5	24
Sugarcane	5	11

Source : Bisalaiah (1990).

Comprehension of growth processes and of interaction between them and the physical-cum-socio-economic environment needs consideration of a large number of dimensions from the historical perspective. Viewed from this requirement, the scope of this paper is quite limited and modest.

Our attempt is to take a look at only two dimensions but these could legitimately be regarded as basic to building up the foundation and the principal supporting structures for the growth process in agriculture. A recurrent theme in development literature relates to the critical importance of technological change in injecting growth impulses into traditional agriculture. This would be even more true of dryland agriculture having an inferior status within the spectrum of traditional agriculture types. Karnataka has a credible record in the field of dryland agriculture research and extension and its impact on production. It would hence be interesting to see how strong are the growth-promoting characteristics of its new technology and how the environment affects them. The next section takes a look at these characteristics in a comparative framework covering traditional dryland, new dryland and irrigated technologies. The theme covered in the third section relates to the price-cost balance reflecting the nature of market environment in which dryland agriculture operates. While the farm lobby has been very vocal in recent years about the unfavourable market environment and about the inadequacy of government intervention to help farmers, it is well known that the anxiety on this score is shared by many other groups, including development economists, free from sectional vested interests and untainted by agricultural fundamentalism. The last section of the paper brings together the implications for research and policy suggested by our analysis.

A word about the data base of the paper. Since the late seventies, the government of Karnataka has been bringing out an annual report on "Region-wise Cost of Cultivation" containing estimates of total cost, its components and returns for the major crops and zones and for the State as a whole. This paper draws on the reports for the years 1980-81 to 1988-89 (Government of Karnataka, 1980-81 to 1988-89). The section on technology is based on the State-level estimates for the major crops for this period and that on price-cost balance uses the zone-wise data for the period 1982-83 to 1985-86. An interesting point to keep in mind while looking at our findings is that the Karnataka reports use a comprehensive cost concept which the farm lobby should find quite acceptable. The wage rate used for computing labour cost (hired plus family labour) is the higher of the prevailing rate and the minimum fixed under the minimum Wages Act, 1948. The cost also includes a margin for risk—10 per cent of total variable cost in the case of irrigated crops and 15 per cent in the case of dryland crops—and a provision for managerial cost based on the size of farm, extent to irrigation and duration of crops. These provisions are in addition to the usual imputations included in the costs to cover owned inputs, capital and land. Our intention is not to endorse the comprehensive concept of cost used in the Karnataka reports but only to suggest that the findings based on it would-be robust in the sense of continuing to hold when an alternative, and less comprehensive, concept of cost is used.

II. TECHNOLOGY

The thrust of a new technology in promoting growth depends on a number of critical characteristics and their combination in a balanced and viable package. The following are the characteristics relevant when assessing the needs of dryland agriculture. First, the technology should have an adequate margin of gross returns over total costs to ensure that the farmer has both economic incentive and investible surplus for adoption of new technology and speedy attainment of its use at the optimum level. Second, the farmers perception of the margin and his degree of confidence in obtaining it steadily over a sufficiently long period depend on the variability in returns over time, particularly the types of variability over which the farmer has little direct control. The characteristics of variability assumes special significance in the case of dryland farmers who are exposed to recurring and ruinous fluctuations. Such fluctuations led to mass exodus from villages in the historical past on a scale which may seem improbable in the prevailing context. However, they are still important enough to affect the farmer's willingness to take risks and his ability to build up immunity to their affect. Third, an adequate margin of returns over costs needs to be achieved through bringing about increases in productivity of land. While this is obvious in the case of land-based activity like agriculture, the constraint distinguishing dryland agriculture is that is has no access to strategies for intensive use of land available to irrigated agriculture. In the case of the former, the increases in land productivity have to be achieved by *conserving* soil and moisture *and* by *choosing* crops/crop varieties having the potential for high productivity under dryland conditions. The realised increases in land productivity would, thus, depend on the researcher's efforts to evolve crops with appropriate characteristics *and* the dryland communities' success in developing adequate arrangements for conservation of soil and water resources; dryland agriculture cannot grow by relying only on 'miracle' seeds or crops. Fourth, the 'conservation' route to increase land productivity needs skilled cultivator practices and careful husbandry. The dryland technology can be—one may even say, ought to be—a genuinely labour-absorbing technology, i.e. a technology which absorbs labour and, simultaneously, provides scope for increase in its productivity. Use of modern inputs like fertiliser would be essential in such a technology but its emphasis would have to be on judicious use—rather than on plentiful—of these scarce inputs whose real costs to the economy often remain concealed because of heavy and uneconomical subsidies.

The characteristics of technology noted above provide some simple yardsticks for assessing the growth potential of new dryland technology in Karnataka agriculture. Our approach is that of a social science researcher and not of a soil scientist or crop specialist. It makes use of the fact that Karnataka agriculture has a wide spectrum of technologies ranging from traditional dryland crops to intensively irrigated crops. The costs, returns, input use and input productivity of these crops clearly reflect the three distinct tiers in the technology in Karnataka agriculture—traditional dryland technology (TD), new dryland technology (ND) and intensive irrigation (II) technology. On a hunch when crops were arranged in the ascending order of gross returns per hectare, we found that the three groups of crops—the lowest third, the middle third and the highest third—could easily be identified with the three tiers of technology. This enables comparison of characteristics of new dryland technology with those of the other two tiers. While the detailed crop-wise tables are given in the Appendix, the findings presented in this section are based on the average (median) characteristics of crops in each of the three groups. Our main interest is in group-level comparisons as the consideration of details of individual crops will take one from the broad economic perspective on the technology to the specific techno-agronomic perspectives appropriate for individual crops. The analysis based on the techno-agronomic perspectives is best left to the agriculture scientist or those in the mainstream of agriculture research.

Tables 7.3 to 7.5 present the comparative technological characteristics of the three groups of crops. The following comments are based on the systematic differences between the three groups noticed in these tables. Table 7.3 covers the core characteristics determining the growth potential of a technology. For convenience in description, the three technologies are referred to as TD (i.e., traditional dryland),

ND (new dryland) and II (Intensive Irrigation) technologies. It can be seen from Table 7.3 that there are striking differences between TD and II. The latter has total cost per hectare nearly six times and gross returns per hectare over eight times of the corresponding levels in TD. What needs to be noted is that intensification of land use implied by the much higher level of total costs per hectare in the case of II as compared to TD is accompanied by more than proportionate growth in gross returns per hectare leading to very impressive margin of net returns. As the last row in Table 7.3 shows, that II technology has positive net returns (i.e. gross returns exceeding the comprehensively defined costs) in 93 per cent of crop-years. A convenient way of contrasting the TD and II technologies is to identify the former with low-productivity poverty-perpetuating technology and the latter with high productivity growth-promoting technology. In sharp contrast to the II technology, the TD technology has very modest net returns on average and, considering the period of nine years covered in our analysis, the net returns were negative in the case of TD crops for a little over half of the crop-years.

TABLE 7.3 : Costs and Returns (Per Hector)

	Technology		
	TD	*ND*	*II*
Total costs (Rs.)	1032	2120	6000
Gross returns (Rs.)	1056	2191	9037
Net returns (Rs.)	24	71	3037
Per cent net returns to total cost	2.4	3.3	50.5
Per cent of years with positive net returns	49	67	93

TABLE 7.4 : Index of Fluctuations

	Technology		
	TD	*ND*	*II*
Total cost	65	55	49
Physical yield	87	71	58
Gross returns	110	74	60
Net returns	405	964	189

Note :

TD = Traditional Dryland

ND = New Dryland

II = Intensive Irrigation

The TD and II technologies provide convenient reference points to locate the ND technology within the range spanning the two extremes of poverty-perpetuating technology at one end and the strong growth-promoting technology at the other. It is obvious from Table 7.3 that while ND is an improvement over TD the two remain close to each other and have a striking family resemblance. Equally striking is the distinct superiority of II over TD and ND placing the former at a much higher tier in the hierarchy of technologies.

TABLE 7.5 : Input Levels And Gross Productivity (Per Hectare)

	Technology		
	TD	*ND*	*II*
Human Labour (days)	46	76	230
Gross returns per day (Rs.)	23	29	39
Modern inputs (Rs.)	79	307	1358
Gross returns per rupee (Rs.)	13.4	7.1	6.7

Note :

TD = Traditional Dryland
ND = New Dryland
II = Intensive Irrigation

These differences provide the key to reading Tables 7.4 and 7.5. Table 7.4 shows the values of index of fluctuations over the period 1980-81 to 1988-89 in the case of per hectare total cost, physical yield, gross returns and net returns. An interesting pattern seen in Table 7.4 is that in each column the index of fluctuations increases as one moves down from total cost to net returns. It is tempting to interpret it thus: total cost reflects the efforts and resources put in by the farmer. Since a substantial part of them are committed in the initial phase of the crop season when little is known about the nature of the full season—favourable of adverse—the farmer has limited degree of freedom in adjusting them to changing conditions. The physical yield, depending as it does on uncontrollable weather factors, has a larger amplitude of fluctuations. Ordinarily, one would expect the weather and the market to counter-balance each other to an extent provided the latter is well organised and behaves in an orderly manner—plausibly, a feature unlikely to prevail in regions of under-development. Finally, the net returns are seen to be the most volatile among the four variables in Table 7.4. This is natural in a residual variable much smaller in magnitude than the two fluctuating variables on which it depends, viz., total costs and gross returns. What is relevant for us to note is the tenuous link between the efforts and resources invested by the farmer (i.e., total costs) and the net returns he obtains because of the intervening uncontrolled forces of weather and market.

Comparison across the three technologies in Table 7.4 shows that the incidence of fluctuations is markedly more in the case of TD that II. More significantly, ND is again seen to be close to TD and as far as fluctuations in net returns are unconcerned, its position is much worse than TD. Considering the modest difference in net returns between the two technologies (see Table 7.3), a reasonable inference from Table 21.4 is that the dryland farmer would have a strong temptation to remain in the TD camp rather than shift over to ND! We have an interesting glimpse here to how the context of underdevelopment inhibits the farmer.

Table 7.5 compares the three technologies in relation to absorption of labour and modern inputs. As the intensity of inputs needs to be compared across the technologies along with the corresponding input productivities, data on gross returns per labour day and per rupee of modern inputs are also included in Table 7.5. While the ND farmer does absorb more labour and more modern inputs per hectare than the TD farmer, there is no clear evidence that he is able to do this along with *increase* in their productivity. As regards labour, there is a modest rise in productivity in ND but at an intensity of labour use much less than in II technology. Regarding modern inputs, the nearly four-fold increase in modern inputs per hectare as between TD and ND is accompanied by a fall in their productivity to a level which II is able to sustain at a very much higher intensity of modern input use. On the whole, to go by the indications provided by Table 7.5, ND does not have a credible performance in the matter of improving land productivity (see Table 7.3) through more intensive and judicious use of labour and modern inputs.

The interesting question to raise at this point is: Is the very modest improved performance of ND as compared to TD and its noticeably weak thrust for growth as compared to II due to the inherently low potential of ND or is it due to non-realisation of its full potential at the farm level? Strictly speaking, the answer to this question needs comparison of performance of ND with its own inherent potential. Comparison with the performance of II helps up to a point but it cannot provide any direct evidence on the gap between the actual and the potential for the ND technology. The analysis of gap needs comparisons at the level of individual crops by experts in the field of dryland technologies. While this is not quite a task within our reach, it has been possible for us to put together a number of suggestive clues indicative of a substantial ND gap between the actual and the potential. These clues are presented below.

It was noted above that the success of ND in raising land productivity depends on comprehensive watershed-wise measures for soil and moisture conservation. A recent study of watershed development prgoramme in Karnataka indicated that "nearly two-thirds of the amount was spent on the arable land. This percentage was the highest in the lowest rainfall zone where the integrated approach covering both arable and non-arable land would be most needed . . . during the first three years [of programme], only 6 per cent of non-arable land received treatment as against 31 per cent of arable land . . . while about two-thirds of the arable land are likely to be covered during [the full] period, the coverage of non-arable land would remain negligible at the level of around 15 per cent. The most likely factors for this marked imbalance in the activities are (i) inadequate preparatory planning for the watershed as a whole and (ii) lack of enthusiasm among the people in the efforts to improve the common property land (Rao and Erappa, 1990). As regards the researchers' efforts to evolve crops with characteristics appropriate to the dryland environment, an authoritative document prepared on eve of Eighth Plan by an expert group observes in relation to the dryland areas in Karnataka that "evolving suitable high-yielding rabi Jowar varieties [is] of urgent need . . . suitable high-yielding varieties of green gram, cow pea, black gram and bengal gram as well as resistant to pod borer [need] to be evolved" (University of Agricultural Science, 1989).

While the two assessments given above relate to the ND potential yet to be created, we also get indications that there are impediments in the way of the potential already created reaching the farmer. For example, studies done by agricultural scientists bring out wide gaps (Govila, 1990):

Maize: "An analysis of yield in on-farm demonstrations (33.47 q/ha) and the national average 11-12 q/ha shows a yield gap of about 22 q/ha."

Bajra: "In on-station trials, productivity of HYVs ranged between 1,324 and 2,914 kg per hectare in 1980-87. The HYVs have 2.56 to 6.48 times more grain yield potential than the national average of 450 kg per hectare (1966-1987). Productivity at the national level can be doubled."

It is instructive to contrast the existence of these gaps with the performance of the extension agencies. According to a recent press report, the vice-chancellor of the University of Agricultural Sciences referred in his address to a training programme for farmers to the increase of 40 per cent in ragi yield to 900 kg per acre achieved in the demonstration fields with "just a passing shower and through the application of minimal fertilisers" but added "however, due to poor extension work, which was the responsibility of the states, hardly 30 per cent of the modern agricultural practices developed by agricultural universities and research institutions was being popularised among farmers" (Krishnamurthy, 1990).

The limited purpose of this section has been to describe the manner in which the depressive features typical of an under-development environment impede emergence of new technologies for dryland agriculture. Our findings indicate that dryland agriculture in Karnataka which had a breakthrough in the mid-fifties did not quite succeed in acquiring an adequately growth-promoting technology. There is evidence of the emergence of such a technology in the costs-cum-returns data on the new dryland crop varieties. But comparison with the traditional dryland and intensive irrigation technologies show that after being in the field for a period of over three decades, the new dryland technology still remains close to the traditional technology in its growth-promoting characteristics with a wide gap separating the two from the intensive irrigation technology. The principal depressive features appear to be the following. First, the fluctuations arising as a combined outcome of adverse physical and economic environment affect farmers willingness and ability to adopt the new technology. More important, the weak policy environment leads to very inadequate efforts to accomplish the critical tasks of conservation of soil and water resources, research to evolve new varieties and extension programmes to reach them to farmers. The weak policy environment, in turn, could be due to the policy maker's preference to rely on the more productive parts of Indian agriculture for quick and substantial increases in production. Thus, it is possible to see the influence on the new dryland technology in Karnataka of all the three features of underdevelopment environment mentioned in the first section of this paper.

III. PRICE-COST BALANCE

The term 'price-cost balance' serves to remind us that market forces—reflecting supply and demand conditions—along with technology influence the levels of prices and costs and, through them, determine the economic viability of an activity—in our case dryland agriculture. The manner in which the 'underdevelopment' features of dryland regions inhibit technology has been described in the preceding section. The same features would also have the effect of weakening the market environment and even making it hostile to dryland agriculture. This section present illustrative evidences based on the data provided by the Cost of Cultivation Reports. It would be seen from the Appendix that the valuation and imputation procedures followed in Cost of Cultivation Reports use *post-harvest prices prevailing in the village* for valuation of output and *prevailing village prices or actual prices paid* in the case of cost components (other than those which, in the absence of identifiable market prices, are handled by specifying arbitrary but reasonable 'prices' like interest rate on working capital or rental rate for owned land). An implication of these procedures is that 'net returns' (i.e., the difference between the gross returns and total costs) per hectare referred to in the preceeding section becomes a summary measure of balance between the post-harvest village prices of crops (i.e., 'prices received' for most of the farmers in dryland communities) and the cost of cultivation depending on the prices paid by them.

The Cost of Cultivation Report present data for crops by ten zones in Karnataka (out of which five zones are dry zones). The comparison of gross returns with total costs for a given crop in a given zone and year provides one observation on 'price-cost balance'. Our interest in this section is not in averaging these observations but in obtaining meaningful distribution of observations spread over different crops, zones and years in the hope of getting interpretable patterns. Owning to constraints of time and resources, we could only use the report for four years—1982-83 to 1985-86—which provided a total of 298 observations. The table given in this section are based on the analysis of these observations.

Table 7.6 shows, major crop-wise, the proportion of observation having price less than cost. The crops are arranged in the table in the ascending order of this proportion. The crops seem to fall in three groups. The first four crops—cotton, groundnut, paddy and jowar—have two-thirds or more of the observations with price exceeding cost. It would be reasonable to regard these crops as having favourable market conditions though this does not mean that the growers of these crops are prosperous or that they test 'fair' returns in the market. 'Favourable market conditions' only imply the absence of a persistent weakness in prices in covering costs. In case of the next four crops, the market conditions cannot clearly be characterised as being either 'favourable' or 'unfavourable'. The growers of such crops operate in a setting with very limited visibility of emerging conditions and hence are likely to prize security above innovation and expansion. The last crop or the last two crops in the table—both coarse grains—are clearly crops with depressed markets. Why do farmers continue to grow such corps? The likeliest answer is that they are grown for self-consumption, in which case the whole question of price-cost balance becomes somewhat irrelevant though, in stricts logic, the farmers ought to make better choices than grow such crops. Equally plausibly, the farmers seek an optimum combination of crops, i.e. crop diversification, which may provide room for inclusion of such crops. Our interpretation of Table 7.6 is confined to taking note of the three fairly distinctive groups of crops noticeable in the table and, more importantly, pointing out the hint provided by the table that five out of the nine crops in the table (i.e. a majority) have market conditions not conducive to growth. Such conditions appear to be pervasive, rather than peripheral, in their prevalence in dryland agriculture.

TABLE 7.6 : Cropwise Observations Having Price Less Than Cost

Crop	*Total Number of Observations*	*Percent having Price less than Cost*
Cotton	23	17
Groundnut	46	21
Paddy	88	26
Jowar	57	34
Sugarcane	18	44
Tobacco	8	50
Tur	6	50
Ragi	37	54
Bajra	15	73
All crops	289	34

It is, hence, important that conceptual frameworks and developed for more detailed analysis of differences between price and cost. We have tried out two approaches, the results of which are reported here, First the difference between price and cost is expressed as per cent of cost. Observations are then distributed in three size classes of the margin of difference :

Size Class of Margin (Per cent)	*Symbol Used in Tables*
0 to 10	**Low Margin (L)**
10 to 20	**Medium Margin (M)**
20 and above	**High Margin (H)**

Similar classes are defined for negative margins, viz., 0 to 10 per cent (Low Margin or L), –10 to –20 per cent (Medium Margin or M) and —20 per cent and below (High Margin or H). The crop-wise distribution of observations by the size of margin is presented in Table 7.7. In the case of cotton and groundnut, the 'High' positive margin class is seen to be the modal class with respectively, 70 per cent and 59 per cent of total observations. These crops also have negligible proportion of observations in the 'High' negative class. It is possible that the growers of these commercial crops have better capacity to adjust to markets and make more effective use of the opportunities for gain they provide. As regards the next two crops which are foodgrains— viz., paddy and jowar—they are marked by a perceptible fall in the frequency in 'High' positive class as compared to cotton and groundnut and a noticeable increase in the frequency in the 'High' negative class. While sugarcane occupied a low rank—lower than even jowar—in Table 7.7, it appears to have a distribution similar to that of cotton and groundnut. Tobacco is the only commercial crop in the table with a pattern of distribution of observations quite different from that of other commercial crop (see Table 7.15 for an interesting clue).

TABLE 7.7 : Distribution of Observations by Margin

Crop	*Margin*						
	Positive			*Negative*			*Total*
	H	*M*	*L*	*L*	*M*	*H*	
Cotton	70	4	9	4	9	4	100
Groundnut	59	7	13	6	13	2	100
Paddy	52	7	15	10	6	10	100
Jowar	47	7	12	11	9	14	100
Sugarcane	44	6	6	39	5	—	100
Tobacco	25	—	25	25	—	17	100
Tur	17	17	16	33	—	25	100
Ragi	32	—	14	24	8	22	100
Bajra	—	7	20	13	33	27	100
All crops	47	6	13	14	9	11	100

It is likely that the differences between cotton and groundnut on one hand and paddy and jowar on the other arise from structural differences between the markets for commercial crops and those for foodgrains. However, we would like to suggest the further proposition that the farmers habitually dealing in weak markets would get little opportunity and experience to acquire 'marketing' skills like keeping oneself informed about changing conditions, forming intelligent guesses about the relevant future events and driving hard bargains. This proposition seems to be particularly true of crops like tur, ragi and bajra, having patterns of observations with an even larger component of adverse fluctuations than paddy and jowar. Thus, Table 7.7 indicates two broad types of market situations—strong markets along with farmers having relatively better market skills in the case of commercial crops (excepting tobacco) and weak markets along with farmers having meagre market skills in the case of foodgrain crops. It is true that there are awkward exceptions like tobacco and, further, that there are considerable variations within the group of foodgrains. However, it would be reasonable to conclude from Table 7.7 that a weak market environment (i.e. the, weak markets along with farmers having weak skills) affect a large number of crops in dryland agriculture excepting a few well-developed commercial crops.

The second approach used by us for analysing the observations is of help in considering the impact of fluctuations on farmer behaviour. This approach is based on the following four components of costs which together add up to total costs:

Component	*Symbol*
Local seeds + human labour + bullock labour + manure	a_1
HYV seeds + fertilisers + pesticides + irrigation + repair charges + miscellaneous expenses	a_2
Land revenue and taxes + interest on working capital + depreciation charges + interest on investment + rental value of land	a_3
Managerial cost + risk premium	a_4
Total Cost	$a_1 + a_2 + a_3 + a_4$

Based on these components, we define the following Price-Cost Balance (PCB) Codes:

Code	*Components of Costs Covered by Price*
1	All the four components, viz, a_1, a_2, a_3 and a_4 are covered
2	a_1, a_2 and a_3 fully covered and a_4 partially covered
3	a_1 and a_2 fully covered, a_3 partially covered and a_4 left fully uncovered
4	a_1 fully covered a_2 partially covered and a_3 and a_4 left fully uncovered
5	a_1 partially covered and all other components left fully uncovered

The hierarchy of cost components implied in these codes can be explained thus. Starting with the worst scenario, viz, a_5, a farmer persistently finding himself in this PCB situation would sooner than later leave farming and become a labourer or migrate elsewhere for employment. If he is in situation a_4, he has still an incentive to continue in farming but economise on 'modern' inputs like HYV seeds, fertilisers, etc. It is clear that a farmer who remains in situation a_4 for long will eventually get into situation a_5 and, thus, move out of farming. Compared to a_4 and, a_5, situations a_3 and a_2 have different behavioural implications regarding what the farmer would do in these situations. He is unlikely to leave farming altogether unless he has a distinctly better alternative activity to take up. At the same time, it is equally unlikely that he would have the incentive or ability to maintain and improve his 'fixed' capital, viz, land, implements, etc. He would be typical of poor farmers who cling on to their activity with little prospects of becoming viable. Lastly, the situation a_1 where all cost components are covered is the best situation in our scheme though it is only fair to mention here that in dryland agriculture there would be severe constraints on the levels of productivity and affluence attainable by the farmers in situation a_1. In fact, they would have to be ever vigilant to remain where they are without slipping into the non-viable situations mentioned above (viz, a_2 to a_5).

It can be seen from Table 7.8 that, taking all the crops together, two-thirds of our observations related to farmers in situation a_1 (viable but not necessarily very productive or affluent) and a little less than another third of total observations related to farmers who are clinging on to land without being fully viable. The other two situations—a_4 and a_5—can be regarded as being quite exceptional and rare though they do have a far-from insignificant presence in the case of two crops, tobacco and ragi. Unlike the distribution by margins (Table 7.7), the distribution by PCB Codes might be more volatile over time. Hence, it would be prudent not to read too much into the inter-crop differences seen in Table 7.8. For example, both sugarcane and tur—very dissimilar crops—have similar distribution in Table 7.8 with all observations falling in situation a_1 or a_2 and with each accounting for a half of them.

The main implication of Table 7.8 is that, along with a strong programme for technological change, dryland agriculture would also need strong price support measure and credit delivery system. Here, again, it is found that the present arrangements for support are very

TABLE 7.8 : Distribution of Observation by PCB Codes

Crop	*1*	*2*	*3*	*4*	*5*	*Total*
Cotton	83	9	8	—	—	100
Groundnut	79	9	12	—	—	100
Paddy	74	16	10	—	—	100
Jowar	66	19	11	—	4	100
Sugarcane	56	44	—	—	—	100
Tobacco	50	25	13	12	—	100
Tur	50	50	—	—	—	100
Ragi	46	30	8	11	5	100
Bajra	27	47	26	—	—	100
All crops	66	21	10	2	1	100

weak. We quote from the observations made by two experts having long experience in these areas. D. S. Tyagi, Member-secretary of the Commission for Agricultural Costs and Prices, offers the following assessment of price support measures: "(a) major failure of the food policy has been in terms of assuring reasonable prices to coarse grain producers. During the peak harvesting season, in years of excess supply, the prices of coarse cereals have ruled below the minimum support price in many markets. The most important reason for this failure in the implementation of the price policy in the case of coarse cereals was the absence of an assured outlet for their final disposal" (Tyagi, 1990). In simple words, the policy-maker fixed minimum support prices for coarse cereals but—like the minimum wages for agricultural labourer—forgot to make arrangements for enforcing them! It is difficult to think of a more revealing glimpse of the policy-maker's attitude to dryland agriculture.

Regarding the credit delivery system, Kahlon on the basis of his careful review of NABARD studies and experiences in the field comes to the conclusion that "it is highly important that the defective loan policies and procedures of lending institutions which impair farmer's ability to secure adequate incremental returns to be able to repay his loans, engage more serious attention than has been done in the past... Studies show that defaults in loan repayments in many cases are involuntary or 'non-wilful' arising from lending banks' imperfect perception of farmers' repaying capacity or inflexible procedures, investment failures from faulty lending/recovery procedures followed by credit institutions and other circumstances beyond the control of borrowers. Some of the major deficiencies in the lending procedures and policies of the credit institutions which contribute to involuntary default of the borrowers, are : under financing, short loan maturities, absence of initial grace period, delays in sanction and disbursement of loans and failure of investments, or investments remaining incomplete due to under-financing" (Kahlon, 1991). While the concern of Kahlon is with the effectiveness of the institutional rural credit system as a whole in India, it is easy to see that the defects he observes are likely to be even more severe in dryland agriculture areas. This would also be true of the obstacles they put in the way of processes of technological change going on in dryland agriculture.

Apart from adding thrust to agricultural growth in Karnataka, the new dryland technology has also a further important role to play. Karnataka has currently about 20 per cent of cultivated area under irrigation. Irrigation is a high priority programme in the State. The eventual irrigation potential in the State is estimated to be around 40 per cent of cultivated area. When the full irrigation potential is developed, its spatial spread is likely to be uneven leaving many areas out of its benefits. In the absence of a new dryland technology, the spread of irrigation is likely to further darken the shadow of under-development it casts on the left-out areas. In principle, such areas could take up other non-agricultural activities but, in practice, the strategy for rural diversification is likely to turn out to be an even harder nut to crack. This is not the place to go into these larger strategy questions. However, the data available in the Cost of Cultivation Reports do provide suggestive clues to the potential thread posed by irrigation to dryland agriculture. Tables 7.9 to 7.15 present separately for the major crops the distribution of observations by margin and by PCB Codes for their important variants. As some of these distributions are based on small numbers of observations, we are only drawing the following very tentative conclusions from these tables.

(i) Cotton (Table 7.12) appear to be the only crop having a new dryland technology performing better than the irrigation technology.

(ii) In the case of Jowar (Table 7.9), the new dryland technology has a good record but, at the moment, would find it very difficult to compete with the irrigation technology;

(iii) As regards ragi and groundnut, no credible alternative to irrigation technology is dissernible in our data (Tables 7.10 and 7.13);

(iv) It is possible that the period of four years considered in our analysis has been exceptionally adverse for bajra. If we go by Table 7.11, bajra has no dependable technology at all as both irrigation technology and new dryland technology are seen to be performing about equally poorly; and

(v) Paddy is an irrigated crop or is grown in assured rainfall areas. What is interesting about this crop is that, even under this favourable circumstance, the performance of rainfed HYV is distinctly inferior to that of irrigated HYV. As a layman would see it, paddy appears to indicate that the task of upgrading the technology for rainfed agriculture is difficult even in assured rainfall areas. It follows that such difficulties would be far more formidable in areas with low and uncertain rainfall where the dryland crops are grown.

TABLE 7.9 : Detailed Table for Jowar

Crop	*Margin*						*Total*						*Total*
	Positive			*Negative*				*PCB Codes*					
	H	*M*	*L*	*L*	*M*	*H*		*1*	*2*	*3*	*4*	*5*	
Kharif, dry, local	14	7	7	29	7	36	100	21	36	29	—	14	100
Kharif, dry, high-yielding variety	60	—	10	—	20	10	100	20	2	10	—	—	100
Kharif, irrigated, high-yielding variety	100	—	—	—	—	—	100	100	—	—	—	—	100

TABLE 7.10 : Detailed Table for Ragi

Crop	*Margin*						*Total*						*Total*
	Positive			*Negative*				*PCB Codes*					
	H	*M*	*L*	*L*	*M*	*H*		*1*	*2*	*3*	*4*	*5*	
Kharif, dry, local	—	—	20	20	10	50	100	20	20	30	30	—	100
Kharif, dry, high-yielding variety	15	—	15	39	16	15	100	31	54	—	8	7	100
Kharif, irrigated, high-yielding variety	90	—	—	10	—	—	100	90	10	—	—	—	100

TABLE 7.11 : Detailed Table for Bajra

Crop	*Margin*						*Total*						*Total*
	Positive			*Negative*				*PCB Codes*					
	H	*M*	*L*	*L*	*M*	*H*		*1*	*2*	*3*	*4*	*5*	
Kharif, dry, local	—	—	20	—	20	60	100	20	40	40	—	—	100
Kharif, dry, high-yielding variety	—	20	20	20	20	20	100	40	20	40	—	—	100
Kharif, irrigated, high-yielding variety	—	—	25	25	50	—	100	25	75	—	—	—	100

TABLE 7.12 : Detailed Table for Cotton

Crop	*Margin*						*Total*						*Total*
	Positive			*Negative*				*PCB Codes*					
	H	*M*	*L*	*L*	*M*	*H*		*1*	*2*	*3*	*4*	*5*	
Kharif, dry, local	56	—	11	—	22	11	100	67	11	22	—	—	100
Kharif, dry, high-yielding variety	80	10	10	—	—	—	100	100	—	—		—	100
Kharif, irrigated, high-yielding variety	75	—	—	25	—	—	100	75	25	—	—	—	100

TABLE 7.13 : Detailed Table for Groundnut

Crop	*Margin*						*Total*						*Total*
	Positive			*Negative*				*PCB Codes*					
	H	*M*	*L*	*L*	*M*	*H*		*1*	*2*	*3*	*4*	*5*	
Kharif, dry	52	6	10	10	16	6	100	71	13	16	—	—	100
Summer, irrigated	67	7	20	—	6	—	100	94	—	6	—	—	100

TABLE 7.14 : Detailed Table for Paddy

Crop	*Margin*						*Total*						*Total*
	Positive			*Negative*				*PCB Codes*					
	H	*M*	*L*	*L*	*M*	*H*		*1*	*2*	*3*	*4*	*5*	
Kharif, dry, local	17	4	22	17	13	27	100	44	39	17	—	—	100
Kharif, dry, high-yielding variety	55	18	18	9	—	—	100	91	9	—	—	—	100
Kharif, irrigated, local	39	15	15	15	8	8	100	69	23	8	—	—	100
Kharif, irrigated, high-yielding variety	95	—	5	—	—	—	100	100	—	—	—	—	100

TABLE 7.15 : Detailed Table for Tobacco

Crop	*Margin*						*Total*						*Total*
	Positive			*Negative*				*PCB Codes*					
	H	*M*	*L*	*L*	*M*	*H*		*1*	*2*	*3*	*4*	*5*	
Dry VFC	—	—	25	14	—	50	100	25	25	25	25	—	100
Dry Bidi	50	—	25	25	—	—	100	75	25	—	—	—	100

Note : An interesting clue given by this table is that the wide spread of tobacco observations in Tables 7.7 and 7.8 arise from the combination of two varieties having markedly different distributions. It may be mentioned here that in the Cost of Cultivation Reports, Dry VFC is reported only in 'Southern Transitional Zone' and Dry Bidi in 'Northern Transitional Zone'. Hence, tobacco is not included in the zone-wise tables (Tables 7.16 to 7.19). VFC Tobacco, which is used for cigarettes, is seen to have a weaker market then Bidi Tobacco.

TABLE 7.16 : Summary Table on Zone-wise Costs Per Unit of Crops

Crop	*Traditional Dryland*		*New Dryland*		*Intensive Irrigation*	
	Cost	*Zone*	*Cost*	*Zone*	*Cost*	*Zone*
Kharif paddy	175	Northern Transitional	160	Southern Transitional	151 (Local) 119 (HYV)	Southern Transitional Central Dry
Rabi paddy	—	—	—	—	283 (Local) 212 (HYV)	Coastal
Summer paddy	—	—	—	—	117	Northern Dry
Kharif jowar	170	Northern Dry	128	Northern Transitional	124	Northern Dry
Rabi jowar	179	North-Eastern Dry	—	—	—	—
Ragi	174	Northern Dry	189	Southern Transitional	121	Southern Transitional
Rainfed cotton	391	North-Eastern Dry	392	Hill	—	—
Groundnut	342	North-Eastern Dry	—	—	238	Eastern Dry
Sugarcane	—	—	—	—	182	Southern Dry

Note : Karnataka has five dryland zones—North-eastern Dry, Northern Dry, Central Dry, Southern Dry and Eastern Dry

The potential competition faced by dryland agriculture also needs to be studied from another, but related, point of view. While in Tables 7.9 to 7.15 the focus is on the potential competition between the dryland technology and irrigation technology within the same crop, one can also look at this issue from the complementary perspective of potential competition between dry zones and the other zones in Karnataka. Owing to limited observations, the zone-wise analysis cannot be done in terms of distribution of observation by margin and PCB Codes. What was feasible to do was to calculate the per unit cost of crops zone-wise. The data on zone-wise cost of crops are presented in Tables 7.17 to 7.20. The only crop variants excluded from Tables 7.17 to 7.20 are those for which observation on cost was available only for one out of four years; crops with 2, 3 or 4 years of observations are included. The summary of findings from Tables 7.17 to 7.20 is given in Table 7.16. A word of explanation about this table. For each crop variant, the table indicates the zone with minimum cost for the variant and the value of the minimum cost in rupees per unit of crop. The implied assumption is that a zone would have a competitive advantage over other zones in growing a crop variant in which it has minimum cost. Cost is only one of the factors as competitiveness of a zone would also depend on the extent of suitable land, availability of irrigation and competition from other corps for these resources. Hence, the indications provided by Table 7.16 are to be taken only as a first and tentative guess which needed to be investigated further.

There are two paradoxical findings in Table 7.16. First, the new dryland variants' of crops reach their minimum costs in northern and southern transitional zones and in hill zone! Second, and equally in congruously, the central, southern and eastern dry zones have competitive advantage only in three irrigated crops—respectively, paddy, sugarcane and summer irrigated groundnut! It is easy to see the precarious position of agriculture in these dry zones, particularly in those parts of these zones which would remain left out even after the full irrigation potential in Karnataka is utilised. Without a viable new dryland technology, their growth prospects would indeed be grim. Compared to the central, southern and eastern dry zones, the northeastern and northern dry zones have competitive advantage in a number of traditional dryland technology crops.

TABLE 7.17 : Zonewise Costs Per Unit of Sugarcane, Cotton and Groundnut

	Sugarcane	*Rainfed Cotton*		*Groundnut*	
		Local	*HYV*	*Kharif Rainfed*	*Summar Irrigated*
Coastal	—	—	—	—	—
Hill	—	—	392	—	—
N E T	—	472	—	404	—
N T	210	—	502	364	—
S T	191	—	—	376	339
N E Dry	—	391	—	342	349
Northern Dry	216	455	439	451	507
Central Dry	—	—	450	449	—
Southern Dry	182	—	—	411	336
Eastern Dry	—	—	—	—	238

TABLE 7.18 : Zonewise Costs Per Unit of Ragi

(Rs)

Zone	*Ragi (Kharif)*		
	Rainfed Local	*Rainfed HYV*	*Irrigated HYV*
Coastal	—	—	—
Hill	—	—	—
N E T	—	—	—
N T	—	—	—
S T	233	189	121
N E Dry	—	—	—
Northern Dry	174	—	—
Central Dry	300	244	—
Southern Dry	295	207	161
Eastern Dry	—	—	151

TABLE 7.19 : Zonewise Costs Per Unit of Jowar

(Rs)

Zone	*Kharif*		*Rabi*	
	Rainfed Local	*Rainfed HYV*	*Irrigated HYV*	*Rainfed Local*
Coastal	—	—	—	—
Hill	—	—	—	—
N E T	—	147	—	226
N T	—	128	—	184
S T	250	—	—	284
N E Dry	—	—	—	179
Northern Dry	170	154	124	210
Central Dry	250	162	—	—
Southern Dry	258	172	—	—
Eastern Dry	—	—	—	—

TABLE 7.20 : Zonewise Costs Per Unit of Paddy

Zone	*Kharif*				*Rabi*		*Summer*
	Rainfed Local	*Rainfed HYV*	*Irrigated Local*	*Irrigated HYV*	*Irrigated Local*	*Irrigated HYV*	*Irrigated HYV*
Coastal	245	164	—	—	283	212	211
Hill	186	164	—	—	—	—	—
N E T	242	—	—	—	—	—	—
N T	175	—	—	—	—	—	—
S T	267	160	151	143	—	—	—
N E Dry	227	—	—	—	—	—	—
Northern Dry	—	—	—	130	—	—	117
Central Dry	—	—	—	119	—	—	125
Southern Dry	—	—	162	130	—	—	133
Eastern Dry	—	—	249	130	—	—	129

It is important to take note of an implication for future suggested by Table 7.16. Full utilisation of irrigation potential in Karnataka, which is likely to be achieved over the next two or three decades, and simultaneous emergence and spread of viable dryland technologies are likely to bring about major shifts in comparative advantage of different zones and long-term growth prospects of different crops. It would be necessary to foresee these shifts in advance and to help dryland zones—and the communities in them—to adjust to these shifts. With growing commercialisation of agriculture and its integration with the larger economy, these communities would find it increasingly difficult to carry on as they have done in the past, by relying on survival strategies in an agriculturally precarious environment. The pace change is likely to be rapid in the future and the price for inability to adjust far more heavy than in the past. Dryland agriculture seems poised to enter a new phase where it has to choose between viable adjustments on one hand and disintegration on the other without having the option of continuing indefinitely in its current status of uneventful stagnation after a modest spurt in growth.

IV. CONCLUSIONS

The main theme of this paper has been that in recent years the forces of under-development have overwhelmed the impulses for growth generated in Karnataka agriculture during the period the mid-fifties to the mid-seventies. In terms of data presented in section II and III, these forces have seriously impaired the thrust for growth of the new dryland technology and weakened the market environment to make it is serious obstacle to growth. This phase in Karnataka's agriculture needs detailed probing to trace the sequence of policies, programmes and related events which led to this outcome. Our tentative guess is that lack of adequate priority for irrigation extension and neglect of the growth potential of dryland agriculture—both, probably, arising from the slackening of developmental efforts in agriculture in the wake of 'green revolution' and the resulting easing of pressures on the food front—were the proximate factors. It must be mentioned here that, at the all-India level, we have a number of assessments for the eighties pointing out the weaknesses in our agricultural performance (CAPP, 1990; Sawant, 1983). They have focused on sluggishness in agricultural investment and growth, lack of adequate attention to degraded lands and indifference towards those parts of agriculture which are the poor cousins of 'green revolution' agriculture.

In a sense, tracing of the specific sequence of events which occurred in Karnataka in somewhat less relevant for the main theme of this paper than may appear at first sight. What we need is a persuasive demonstration of a link between features of 'underdevelopment' on one hand and the persistence of stagnation and weakening of growth processes in dryland agriculture on the other. An argument in an outline is presented below to suggest that a possible linking variable is the poor image which policy making-cum-influencing elites have to the rural people and their resources, particularly in the backward areas. It is not that the poor image always arises from any animus against the poor or a desire to exploit them. In fact, a thought which frightens us is that the image is possibly shared even by those who have a genuine sympathy for the poor and urge to help them. It lead to strategy and approaches based on lack of faith in the latent potential of the poor people and poor resources to contribute to development. Not surprisingly, they result in schemes many of which remain ill-planned, half-hearted and poorly-implemented despite full awareness on the part of the policy-maker and the implementer of these defects. Rolling back underdevelopment is a task at the level of reconstruction of society which no government anywhere—much less one in India—can take up and complete with its own efforts alone. However, with a genuine empathy for the poor and rapport with them the government and others in the field of development could certainly do much better than now.

After all, there are an embarrassingly large number of cases—though they remain a speck in the overall map of India—where dedicated and charismatic activists have been able to mobilise disadvantaged groups, helping them to reach profoundly moving heights of self-assertion and self-effort.

We conclude the paper with a few easily recognisable illustration of negative perceptions about the poor people and poor resources held by the 'mainstream' groups making and influencing policies.

(i) After over two decades of official and academic exercises to measure poverty, the poverty-line still remains tied up with the 'minimal' concept based on meeting the calorie needs. It is certainly not beyond our ingenuity to define a power-line with adequate allowances for education, nutrition, health, housing, etc. Would it be a carping comment to say that the 'minimal' poverty line reflects an approach to poverty removal remaining confined to elimination of mass hunger and starvation. Dimensions other than open hunger would get low priority in this approach and, predictably, that is precisely what seems to have happened to schemes for primary education, primary health etc. with far more modest achievements than what one would expect from confessional declarations in the assessment of each Plan to do much better in the next one.

(ii) What is the potential productivity of the poor's abundant resource, viz. his labour? The recent and continuing debate on evaluation of labour input while computing farm costs provides an interesting glimpse of the point of view of experts. It is indeed reasonable to say that if an employer pays farm labourers a wage rate lower than the minimum rate specified by the Act, then the actual wages paid should be included in the costs and not the wages computed by using the minimum rate. As the choice is obvious when the actual wage rate exceeds the minimum rate, it follows that the wage labour component of farm costs should be based on the actual rate and not the minimum rate. Eminent experts have expressed serious reservations on the government's recent decisions to take the higher among the two rates for the costing of labour. It is interesting to consider the implications of the stand taken by experts in the case of dryland agriculture. The typical situation in dryland agriculture is the one in which most of the labour on farms is provided by the family itself and the labour market remains depressed with low and erratically-moving wage rates. How should the family labour—i.e. the major part of the labour component—be evaluated in this situation when deciding on minimum support and procurement prices? Based on sound economic logic, the experts would insist on using the prevailing rate which, very likely, would be lower than the minimum rate. However, computation of minimum support and procurement prices using the prevailing wage rate means, in effect, that the government itself would pay the farmer through price-fixation an implied wage rate lower than the minimum rate! Would such a procedure not mean that the government as well as the experts advising it value labour at a much lower rate than the floor level—i.e, the minimum wage rate—which they themselves have specified for payment to be made to labour? Would it be either valid or fair to input low productivity to family labour when, in terms of care, diligence and skills, it is likely to be superior to unskilled wage labour? A little reflection would show that the objections to the inclusion of management costs and risk margins while computing the minimum support and procurement prices would also be subject to the same doubts. With schemes for crop insurance bobbing up now and then with little prospect of becoming a strong and stable component in the support system for agriculture, it would indeed seem unfair to poor farmers to fix the minimum support price—which become operative only when market prices are low—without a risk margin. On similar reasoning, there would be a case for including the risk margin in procurement price which is the price paid by the government for that part of the produce acquired from the farmer at lower than market price. The basic question is: If crop insurance is impractical, could be problem which it seeks to tackle be just left for the farmers to handle on their own?

(iii) Regarding poor resources not receiving adequate policy attention, we refer to two items which are particularly relevant to mention in the case of Karnataka. Karnataka has a long tradition of tank irrigation which has been maintained by the rural communities' own efforts. Tanks are now 'dying' at an alarming rate with no alternative irrigation source in sight. Experts take the view that revival of tanks is uneconomical which only means that it is far too expensive to revive old tanks—or construct new ones—through government schemes farmed out to contractors (Rao and Chandrakant, 1984). Would not the situation look different if people are mobilised—not for free labour as in the historical past but through employment programmes paying minimum wages? As a second example, we refer to the non-arable and common property lands which get neglected, as was mentioned in the section on technology, even in watershed development programmes. Jodha has pointed out that "development of watersheds in dry regions need public investment to the tune of Rs. 2,000 per hectare whereas land development in command areas required about Rs. 12,000 to Rs. 15,000 per hectare" (quoted in Reddy, 1989). It is unnecessary to labour the point that a policy-maker looking at a situation like dryland agriculture in Karnataka would have every temptation and a host of impeccable arguments to conclude that the poor people and the resources available in their environment are far too expensive to develop!

APPENDIX-A

The annual reports on "Region-wise Cost of Cultivation" provide data for 33 crops and mixed crops. Data for all the years are not available for each of these crops. We found that there are 18 major crops for which data are available for most of the years in the period 1980-81 to 1988-89. The index of fluctuations given in Table 7.A2 is based on the formula

$$\frac{\begin{array}{cc}\text{Maximum Value} & \text{Minimum Value}\\ \text{(between 1980-81 and} & \text{(between 1980-81}\\ \text{1988-89)} & \text{and 1988-89)}\end{array}}{\text{Median Value for the years 1980-81 to 1988-89}} \times 100$$

The tables given in the section on technology in the text are based on the median of the crop values within each group.

The paragraphs below (reproduced from Region-wise Cost of Cultivation-Reports) describe the valuation and imputation procedures used in computing costs and returns.

The total cost were divided into two broad classes (i) Variable cost (ii) Fixed cost. The method adopted for computing the different cost items are described below:

(i) Variable Cost

(a) *Seed* : Farm produced seed has been evaluated at the village prices prevalent at the time of sowing. Purchased seed has been evaluated at actual rates paid by the sample farmers.

(b) *Farm Yard Manures:* Farm produced manure was evaluated at the village prices prevalent at the time of sowing. FYM purchased was evaluated at actual rates paid by the sample farmers.

(c) *Fertilisers and P P Chemicals :* Costs for fertilisers and plant protection chemicals were the amounts actually paid by the sample farmers.

(d) *Irrigation charges :* In case of canal and tank irrigated crops as the water rate chargeable from time to time under the Karnataka Irrigation (levy of water rate) Rules 1965 was taken into account. In case the crop is irrigated by lift irrigation, the electrical charges or diesel charges as the case may be were determined in proportion to the acreage under the crops.

TABLE 7.A1 : Technological Characteristics of Crops (Per Hectare)

Crops	*No. of Years for which Crop Data Available*	*Total Cost (Rs.)*	*Gross Returns (Rs.)*	*No of Years with Positive Net Return*	*Human Labour Days*	*Modern Inputs (Rs.)*
	1	2	3	4	5	6
		TD				
Bajra (local rainfed, kharif)	8	905.65	775.85	Nil	45.14	12.97
Jowar (local rainfed, rabi)	9	796.28	879.06	8	35.00	55.49
Cotton (local rainfed, kharif)	9	872.12	1014.30	6	44.92	51.99
Jowar (local rainfed, kharif)	9	1158.36	1098.14	3	58.70	102.01
Ragi (local rainfed, kharif)	9	2025.28	1595.35	Nil	90.67	165.16
Bengal gram (rainfed, rabi)	9	1479.45	1770.15	9	46.57	118.56
		ND				
Red gram (rainfed, kharif)	8	1077.07	1208.19	4	29.68	307.31
Jowar (high-yield variety, rainfed, kharif)	8	2083.41	2078.36	4	75.93	327.98
Ragi (high-yielding variety, rainfed, kharif)	9	2119.97	2191.04	5	94.04	271.74
Groundnut (rainfed, kharif)	9	2184.99	2456.68	8	75.30	135.16
Beedi tobacco (rainfed, kharif)	9	3600.00	4488.00	8	228.04	492.06
		II				
Rag (high-yielding variety, irrigated, kharif)	9	3866.00	5020.16	9	164.85	515.99
Groundnut (irrigated, summer)	9	4865.65	6308.24	8	132.18	822.03
Paddy (high-yielding variety, irrigated, rabi)	7	5921.53	7075.00	5	182.00	632.09
Paddy (high-yielding variety, irrigated, summer)	9	5999.95	9037.40	9	229.94	1375.95
Paddy (high-yielding variety, irrigated, summer)	9	6212.95	9397.62	9	237.40	1358.06
Cotton (high-yielding variety, irrigated, kharif)	8	11484.98	14570.72	7	431.96	4486.10
Sugarcane (planting)	9	16536.36	20809.47	9	467.33	4534.56

(e) *Repair charges :* Repair charges were based on the sample farmers estimate of normal annual repairs for each item of equipment used. Repair charges for a particular crop is determined in proportion to the acreage under the crop. In case certain implements/equipments are used only for a particular crop the entire repair charges were charged to that crop only.

TABLE 7.A2 : Index of Fluctuations

Crops	*Total Cost*	*Physical Yield*	*Gross Returns*	*Net Returns*
1	2	3	4	5
	TD			
Bajra (local rainfed, kharif)	166	134	166	233
Jowar (local rainfed, rabi)	55	60	77	364
Cotton (local rainfed, kharif)	104	118	191	522
Jowar (local rainfed, kharif)	31	81	34	449
Ragi (local rainfed, kharif)	67	63	35	2842
Bengal gram (rainfed, rabi)	63	92	144	276
	ND			
Red gram (rainfed kharif)	102	83	154	1040
Jowar (high-yielding variety, rainfed, kharif)	47	71	65	964
Ragi (high-yielding variety, rainfed, kharif)	50	94	74	2358
Groundnut (rainfed, kharif)	55	39	66	315
Beedi tobacco (rainfed, kharif)	89	56	92	199
	II			
Ragi (high-yielding variety, irrigated, kharif)	97	63	110	179
Groundnut (irrigated, summer)	52	78	81	189
Paddy (high-yielding, variety irrigated, rabi)	36	32	46	348
Paddy (high-yielding, variety irrigated, kharif)	33	35	47	104
Paddy (high-yielding variety, irrigated, summer)	44	58	60	153
Cotton (high-yielding variety, irrigated, kharif)	53	92	95	352
Sugarcane (planted)	49	26	47	213

(f) *Miscellaneous charges:* This item includes the expenses incurred by sample farmers which cannot be included in any other item of expenditure. For instance, in case of VFC tobacco the miscellaneous expenses include expenditure incurred on account of tobacco board licence fee, transportation charges incurred while procuring input materials required for crop enterprise, cost of obtaining credit if any, paddy straw used to cover seed beds, twine etc.

(g) *Labour :* Labour was charged at the prevailing wage rates paid per day (eight hours) in that locality for men, women and bullock pair. Family labour was charged at the same rate as hired labour. Wherever, prevailed wage rates were less than the minimum wages fixed by the government from time to time under the Minimum Wage Act, 1948, the minimum wages were taken into account in computing labour cost.

(ii) Fixed cost

(a) *Interest on investments on implements:* Estimated annually at the rate of 10 per cent on half the purchase value of the implements till its existence. The allocation of cost to different crops grown by the sample farmers was done in proportion to the acreage under the crop. In case certain implements (such as pump-sets) are used only for a particular crop/crops, the whole amount has been charged to that crop/crops.

(b) *Land revenue and other taxes:* Land revenue and other taxes was charged at the rates levied by the government. Allocation of the cost was done in proportion to the area under the crop.

(c) *Interest on working capital:* Interest has been charged at the rate of 15 per cent per annum for a crop period on the variable cost incurred during the period of cultivation.

(d) *Depreciation charges:* Depreciation charges on implements/ machineries were calculated by straight line method, i.e., by dividing the original cost of item by the expected life of the item as estimated by the sample farmers.

Allocation of depreciation charges to individual crop were done in proportion to the acreage under the crops. In case certain implements/machineries (such as pumpsets) are used only for a particular crop/crops, the entire amount of depreciation charges on account of such implements has been charged to that particular crop/crops.

(e) *Risk cost:* The cost is estimated at 10 per cent on the total variable cost in case of irrigated crops and 15 per cent in case of dryland crops.

(f) *Managerial cost:* Managerial cost is worked out taking into consideration that a farmer can supervise 10 acres of irrigated land as per the ceiling limit and a sum of Rs. 400 is taken as managerial cost per month. This apportioned to an acre works out to Rs. 480 per year and where a crop has duration of six months or less, the managerial cost will be Rs. 240. In case of dryland, one can supervise 50 acres and the managerial cost is apportioned accordingly.

(g) *Rental value:* Rental value has been estimated at 10 per cent of the gross income in case of irrigated crop and 5 per cent in case of dryland. *Procedure for imputation of value of the main product and the by-product:*

The value of the main product and by-product are imputed at the post-harvest prices prevailing in the selected villages.

Procedure for working out cost per quintal

The value of by-product has been deducted from the total cost and the remainder is divided from the yield of main product to get the cost per quintal. In case of mixed crops this method could not be followed since in this method the cost of by-product itself accounted for the major portion of the cost of cultivation. Therefore, cost in case of mixed crops was computed in proportion of the share of each to gross income from the crops included in the mixture.

Note : Reproduced from *Economic and Political Weakly,* March 30, 1991 with kind permission of the publisher.

REFERENCES

Atchi Reddy, M. and Sarojini, Y. (1990), 'Economics of Dry Farming: A Study of Cotton Crop in Rayalaseema 1800-1980', Eighth Annual Conference of Andhra Pradesh Economic Association, Hyderabad, January 1990.

Basu, D. N. and Rajagopalan, V. (1990), 'Agro-Climatic Regional Planning : Regional Indicators and Typologies', *Indian Journal of Agricultural Economics,* Vol. XLV, No. 3, July-September 1990.

Bisalaiah, S. (1990), 'Three Epochs of Technology-Extension Nexus in Karnataka Agriculture', paper presented to Southern Regional Seminar on Agricultural Policy, ISEC, Bangalore, November, 1990.

Committee for Agricultural Policies and Programmes (CAPP) (1990). Report of the Committee submitted in July 1990.

Government of Karnataka (1980-81 to 1988-89), Region-wise Cost of Cultivation Reports.

Govila, O. P. (1990), 'New Direction in Millets', *The Hindu Survey of Indian Agriculture,* 1990.

Kahlon, A. S. (1991), 'Institutional Credit and Overdues: Borrowers' Angle', *Economic and Political Weekly,* February 2, 1991.

Krishnamurthy, K. (1990), Report in *Indian Express* of November 12, 1990.

Planning Commission (1989), Documents circulated in the Meeting of Panel of Economists held in July 1989.

Planning Commission (1990), Approach to Eighth Five-Year Plan.

Rao, V. M. and Chandrakant, M. G. (1984), 'Resources at the Margin: A Note on the Karnataka Tank Irrigation Programme', *Economic and Political Weekly* (Review of Agriculture) June 30.

Rao, V. M. and Erappa, S. (1991), 'Development Opportunities in Dryland Agriculture', in *Agricultural Development Policy,* Indian Society of Agricultural Economics, Bombay.

Ratna Reddy, V. (1989), 'Labour Absorption in Agriculture: Need for an Alternative Development Strategy', *Artha Vijnana,* Vol. 31, No. 4, December 1989.

Sawant, S. D. (1983), 'Investigation of the Hypothesis of Declaration in Indian Agriculture', *Indian Journal of Agricultural Economics,* Vol. 38, October-December 1983.

Tyagi, D. S. (1990), *Managing India's Food Economy,* Sage, New Delhi.

University of Agricultural Sciences (1989), 'Agricultural Development Status and Strategy Paper on Zone 10: Southern Plateau and Hills Region' (mimeographed), Bangalore.

8

INTER-SECTORAL TERMS OF TRADE

A. S. KAHLON & D. S. TYAGI

INTRODUCTION

The debate on inter-sectoral terms of trade, of late, has shifted to a more empirical and concrete issue of whether or not unfavourable terms of trade have contributed to the depressed growth of Indian agricultural sector. Whereas Mason (1966), Schultz (1968), Lipton (1977) and some others held the view that farm prices in developing countries, particularly in India, were kept down systematically, Griffin (1974) has held an opposite view, Dantwala (1967, 1976) has also asserted over and again that incentives including those emanating from higher product prices were available to the agricultural sector and Mitra (1977) has concluded that farm prices were pushed disproportionately upward in India. Doubts, have, however, been raised about the assertion that rise in agricultural product prices have not lagged behind the rise in prices of non-agricultural goods.[1]

A close scrutiny of the methodologies adopted for generating the set of views, particularly those relying on estimation of terms of trade, gives the impression that the debate has been marred by considerable confusion arising from inappropriate data and the methodology used in the analysis.

Part I of this paper examines the various methodological issues involved in the estimation of terms of trade and shows how the results get biased in the various studies conducted on this subject. Part II presents details of an alternative methodology for deriving the terms of trade and the results which *inter alia* indicate that terms of trade have moved against agriculture in the recent years.

PART-I

The terms of trade between the two sectors compare prices of exports of one sector with the price of imports from the other sector. This comparison of export and import prices could be termed as 'net barter terms of trade'. If *Px* is the price of exports and *Pm* price of imports, then the net barter terms of trade would be $\frac{Px}{Pm}$. Following Kindleberger's concept of capacity to import[2], expressed as $\frac{Px}{Pm} \times Qx$ where *Qx* is the index of the quantity exported, some research workers have followed it up to estimate income terms of trade.

The requisite information and the data for the estimation of terms of trade between the two sectors of an economy are:

(i) List of products/commodities exchanged between the two sectors;
(ii) Relative share of each item of 'export' in the total exports and also the share of each imported item in the total imports;
(iii) Prices at which the products/commodities were exchanged; and
(iv) Volume of exports and imports.

As the number of commodities/products exchanged between the two sectors would invariably be more than one and since the comparison has to be made over time, an appropriate method of construction of indices of unit value of exports, and imports is necessary. Indices may be given $\left(\frac{Pi\,qi}{Po\,qi}\right.$ Paasche's formula$\left.\right)$ or base year weights $\left(\frac{Pi\,q_o}{P_o\,q_o}\right.$ Laspeyre's formula$\left.\right)$ current weights or a compromise between the two, i.e., Fisher's 'Ideal' index formula based on geometric average of two index numbers. Each formula has its own merits and demerits.[3]

The degree of accuracy of estimates of terms of trade would obviously depend upon : (i) comprehensiveness of the identified pattern of trade; (ii) the reliability of weights used for constructing the price indices in reflecting exactly the composition of trade between the two sectors; (iii) the capacity of the selected price indicators to truly reflect the movements of prices both for exports and imports, (iv) the sensitiveness of the method used for constructing price indices in reflecting the real changes in the prices; and (v) reliability of the indices of volume of exports. In a situation of a two sector analysis, *viz.*, agricultural and non-agricultural, ideally one has to strive at making a comparison between prices received by the agricultural sector and prices paid for its purchases from the non-agricultural sector.

The studies in which either the estimates of terms of trade have been explicitly worked out or have implicitly been inferred from a comparison of relative price changes between the agricultural and the non-agricultural sectors of the Indian economy could be classified into two: (i) Estimating or inferring about the movement of terms of trade from a direct comparison of index number of wholesale prices for such groups or sub-groups as manufacturers and agricultural commodities or foodgrains, (ii) Instead of relying on the weights assigned in the construction of wholesale prices index numbers for combining various commodities into subgroups, groups etc., using weights for different items for deriving the price indices. The work of Ashok Dar (1968), Sethi (1971), Dantwala (1976), Ashok Mitra (1977), and Sidhu and Singh (1979) falls in that first group while that of Thamarajakshi (1969, 1977) and Venketaraman (1979) falls in the second.

Comprehensiveness of Identified Pattern of Trade

It is obvious that unless the pattern of trade is comprehensively identified so as to include all major products exchanged between the two sectors, the analysis based on this would not be very valid. In this regard, studies falling in the first category suffer from the limitation that such groups as manufacturers, for which index number of wholesale prices are available, are inclusive of many such commodities which were not transacted between the two sectors.[5] However, this all-inclusive group of commodities could account for even such items of capital formation which were left out in the second group of studies, at best, only products exchanged for final and intermediate use were identified. Again, in case of products exchanged between the two sectors for final use, these studies relied upon NSS consumption expenditure surveys which provided details only under 12 heads and, in fact, restricted the number of items identified as purchased by agricultural sector for final consumption to only six-edible oil, sugar, salt, clothing fuel and light and non-food items.[6] Further, presumably due to the absence of detailed break-up of the group 'non-food items' and non-availability of this group in wholesale price indices, this important group was left out while constructing the prices paid indices. It would be anybody's guess as to how meaningful the analysis of terms of trade would be, based on such a limited coverage.

Reliability of Weights

The weights assigned to different commodities in the construction of the wholesale price index are on the basis of the total volume of transactions and do not reflect the quantities sold by one sector to the other. Thus they do not truly represent the composition of trade between agricultural and non-agricultural sector which is so important for proper determination of the terms of trade. As a result, the studies falling in the first group suffered from serious limitations arising out of use of inappropriate weights.

Again, in the studies, falling in the second group, the weights attached to different items were not estimated properly owing to omission of items purchased by the agricultural sector for capital formation. Further, in deriving weights for items sold by agricultural sector to non-agricultural sector for final consumption, the NSS consumer expenditure estimates were used which were based on retail prices and did not reflect the prices received by the farmer owing to wide price spreads in different agricultural commodities.[7] The weights derived directly on the basis of final consumption estimates would thus tend to overestimate the share of those items in case of which the difference between final retail prices and the producer prices was large.[8]

Capacity of Price Indicators

Index numbers of wholesale prices have been used as indicators of prices at which commodities were exchanged between the two sectors in almost all the studies. We have discussed elsewhere (Kahlon, Tyagi) the limitations of these index numbers in the analysis of terms of trade and it has been shown (Tyagi, 1979) that they systematically tend to overestimate the changes in prices of agricultural commodities.[9] It was estimated that the extent of overestimation was of the order of 10 to 15 per cent in a period of 10-12 years and that of the under-estimation in prices of manufacturers was 5 per cent.

Indices of Volume of Exports

A review of the methodology used for estimating indices O_{XI}, i.e., quantity exported from the agriculture sectors, shows that while estimating the export of commodities from agricultural sector to non-agricultural for final consumption, the ratio (R_T) of consumption expenditure on agricultural goods (blown up estimate from NSS consumption data) by the non-agricultural sector to the total blown-up consumption expenditure (agricultural goods+non-agricultural goods) for the entire population (agriculture + non-agriculture) was calculated.[10] This ratio (R_T) was then applied to CSO estimates of final consumption expenditure to yield estimate of final consumption of agricultural goods by the non-agricultural sector which in turn was used as estimate of exports from agricultural sector.[11] It is obvious that such a derivation could provide valid estimates only when the NSS blown-up estimate for agricultural products, non-agricultural products and all products carry uniform proportionate relationship with the corresponding CSO estimates.

Table 8.1 gives the estimates of consumption expenditure on agricultural goods and non-agricultural goods based on NSS and CSO estimates. It was apparent that whereas in 1961-62, NSS estimate of consumption expenditure on agricultural products was 130 per cent of the corresponding CSO estimate, in case of non-agricultural products and total consumption expenditure, it was 78.8 per cent and 103.5 per cent respectively.[12] Over the years, the ratio of NSS estimates of consumption expenditure of non-agricultural products to corresponding CSO estimates further declined from 78.8 per cent in 1961-62 to 63.4 per cent in 1970-71 which means the value of R_T was further inflated. These differences would obviously affect the R_T calculated in these studies in a manner which would inflate the consumption expenditure of agricultural commodities by the non-agricultural sector. To obtain the correct value of R_T the NSS estimates were brought in line with the CSO estimates by adjusting (matching) the former estimates of agricultural and non-agricultural products with those of the latter. These adjusted estimates, presented in Table 8.1, show that while uncorrected NSS blown-up estimates yielded R_T value of 19.8 for 1961-62, the NSS estimate corrected for relatively large under estimation of non-agricultural products reduced the value of R_T to 15.8. This difference between the adjusted and unadjusted R_T widened over the

TABLE 8.1 : Difference between NSS and SCO Estimates of Private Final Consumption Expenditure

		Agricultural Commodities			*Non-Agri. Commodities*			*Total Consumption*				
	Year	*Agri. Population*	*Non-Agri. Population*	*Total*	*Agri. Population*	*Non-Agri. Population*	*Total*	*Agri. Population*	*Non-Agri. Population*	*Total*	*Value of R_T*	*Value of R_T Based on A_{djd} Data*
1	2	3	4	5	6	7	8	9	10	11	12	13
1. NSS estimates (blown up) Rs. crore	1961-62	5046.52	2464.78	7511.30	3018.12	1879.18	4897.30	7511.30	4879.41	12408.71	19.8	
2. CSO estimates (Rs. crore)	"							5778.32	6216.98	11995.30		
3. NSS as per cent of CSO	"							129.99	78.77	103.45		15.8
1. NSS estimates (blown up) Rs. crore	1969-70	9929.18	4878.21	14807.39	5561.95	3420.02	8981.97	14807.39	8981.97	23789.36	20.5	
2. CSO estimates (Rs. crore)	"							1359.36	13761.24	27420.60		
3. NSS as per cent of CSO	"							108.40	65.27	86.75		16.4
1. NSS estimates (blown up) Rs. crore	1970-71	10252.10	5084.28	15536.38	5883.86	3709.80	9593.66	15336.38	9593.66	24930.04	20.4	
2. CSO estimates (Rs. crore)	"							14147.32	15143.60	29290.92		
3. NSS as per cent of CSO	"							108.40	63.35	85.11		16.0

years. It is evident that the inflated value of R_T and consequently of Q_X obtained in different studies contributed to over-estimation of exports from the agricultural sector.

Again, the consumption expenditure data of NSS 4th and 5th rounds used in these studies are not comparable with the data from the subsequent rounds owing to the differences in the reference period and method of evaluating the home grown components. Mukherjee (1968) and Chaudhary (1979) have pointed to this non-comparability of estimates. The effect of the use of 4th and 5th round data in various studies has been that while for 1951-52 and 1952-53, (based on 4th and 5th round) the value of R_T worked out at 16.24 and 16.92 respectively, it jumped to 18.93 in 1953-54 and to 19.34 in 1954-55. These ratios, when multiplied with CSO's estimates of total consumption expenditure, gave under estimates of the consumption of agricultural goods by the non-agricultural sector in the base year. Thus while the estimates for the base year got under estimated, those for the terminal year got inflated. In fact, it is a consequence of this distortion that the rate of growth in the marketed surplus of agricultural sector obtained by Thamarajakshi (1969, 1977) and Venkataraman (1979) turned out to be so high.[13]

It is evident that income terms of trade, estimated on the basis of inflated Qt would be faulty. To that extent these studies have faulted. In fact, the concept of income terms of trade, defined as $\frac{Px}{Pm} \times Qt$ can be itself misleading in the context of the two sector model since, based on this formula, the terms of trade (income) of both the sectors could register significant increases, leading to a conclusion that the purchasing power of both the sectors had increased.[14] On the contrary, the concept of terms of trade implies deterioration in the terms of trade of one sector if it has shown some improvement in the other. This is rightly so because it is the ratio of the two. Thus the concept of income terms of trade based only on the volume of exports of one sector is misleading.

Method of Constructing Price Indices

In almost all the studies on the subject, a fixed base index with base year weights was employed for construction of price indices. The difficulty with this method is that with the passage of time the composition of trade for which prices are being measured no longer reflects the trade that is actually taking place. For example, in the purchases of agricultural sector for intermediate use, fertilisers, diesel and electricity had much larger share in the 1970s than that in the 1950s or the early 1960s; similarly sales of wheat from agricultural sector to the non-agricultural sector have a much larger share today than in the early 1960s.[15]

In such a case, when the price increase has been higher for those commodities whose share in the total purchases by the agricultural sector has also increased, the use of Laspeyre's formula would tend to underestimate the extent of rise in the prices paid by the agricultural sector.[16] On the other hand, the prices of those agricultural commodities, whose supplies have increased relatively at a slower pace, have shown relatively larger increase and thus Laspeyre's formula would overestimate the rise in the prices received by the agricultural sector since it does not take into account the fact that the commodities which have risen the most in price have a much smaller share in total sales from the agricultural sector now than in the base year.[17] In any case, use of Laspeyre's formula seems to be quite inappropriate when price indices are worked out for period extending upto 15-20 years.

To conclude, most of the studies cited on the subject suffered from serious limitations on account of limited coverage, use of improper weights, inappropriate price indicators, adoption of incorrect method for estimating volume of exports and the use of a method for constructing price indices which on *a priori* reasoning would underestimate the rise in prices of non-agricultural goods and inflate the rise in prices of agricultural commodities.

Choice of Base and Terminal Years

The choice of base year, i.e., the year for which prices paid and prices received are considered at par, is crucial to the whole analysis of terms of trade. This is more so in a two sector model, particularly, when the prices of one sector display large inter-year variation owing to the fairly inelastic demand and supply situation. By picking up a drought year as the base year (1965-66 or 1966-67 or 1973-74) when prices of agricultural commodities ruled very high, it can be shown that the terms of trade in the following years remained against agriculture; similarly, by taking a base year in which prices of agricultural goods were low, it can be shown that terms of trade have moved in favour of agriculture. Again when reliance is placed on the compound growth rates for drawing inferences about the movement of terms of trade, the choice of initial and terminal years assumes great significance as it is well known that choice of these two makes a great difference to the calculation of compound growth rates. A study according to which the terms of trade moved in favour of agriculture at the rate of 0.51 per cent per annum (and this result was used repeatedly to show that terms to trade had moved in favour of agriculture) contained 1951-52, and 1965-66 as the two terminals. Table 8.2 gives the growth rate of the net barter terms of trade (using the data of the said study) with different terminal years. It would be seen that when 1965-66 was considered as the terminal year, the terms of trade moved in favour of agriculture. But in other combinations, the terms of trade seem to have moved against agriculture.[18] Thus, the choice of base and terminal years can make two great difference to the conclusions regarding the movement of the terms of trade. And a part of the difference in the results of various studies had its origin in the selection of these years.

TABLE 8.2 : Compound Growth Rates of Net Barter Terms of Trade

Sl. No.	*Period*	*Growth Rate*
1	2	3
1.	1951-52 to 1962-63	0.04
2.	1951-52 to 1963-64	–0.06
3.	1953-54 to 1962-63	0.08
4.	1953-54 to 1963-64	–0.06
5.	1951-52 to 1965-66	0.51

Source : Calculated from the data available in Thamarajakshi's (1969) study.

PART-II

Keeping in view the limitations of the methodologies adopted in estimating the terms of trade between agricultural and non-agricultural sector of the Indian economy, an alternative methodology has been developed. This part discussed the details of the methodology used and the results obtained.

Pattern of Trade

The report of the 26th round of NSS (1978) (Consumer Expenditure for Cultivator Households : Rural) gives in great detail the pattern of consumption expenditure of cultivator households. In this report, for the first time, break-up of expenditure has been given for over 90 items and, particularly, details are available for the non-food expenditure under 56 heads. This makes the identification of products purchased by the agricultural sector from the non-agricultural sector more reliable and comprehensive. On the basis of this report and keeping in view the availability of data on prices of these products, 17 items/commodity groups were selected to represent purchases by the agricultural sector for final consumption.[19] The report of the All-India Debt and Investment Survey gives the composition of gross capital expenditure in farm and non-farm business by the cultivator households. The break-up of the total gross capital expenditure is available under 16 heads. Keeping in view the availability of price statistics and information regarding details of items required in the construction activity, 8 items representing purchases by the agricultural sector for the purpose of capital formation were identified.[20]

The National Accounts Statistics (CSO) gives estimates of value of various inputs used by agricultural sector for production. Five items, chemical fertilisers, electricity diesel oil, pesticides and insecticides, and repair and maintenance of fixed capital were considered as items involving purchases. Two more items, oil cakes and salt and medicine for animals, were also purchased by the agricultural sector. Thus in all 32 items were identified to represent the purchases by agricultural sector from the non-agricultural.

The commodities purchased by the non-agricultural sector from the agricultural sector are either for final composition or for intermediate consumption. Cereals and cereal substitutes, pulses, vegetables, fruits and nuts, milk and milk products, meat, fish and egg, and gur are the items purchased for final consumption, whereas oilseed, fibres, sugarcane, tobacco and rubber are the principal raw-materials purchased by the non-agricultural sector. In this way 22 commodities/commodity groups were identified to represent the basket of goods sold by the agricultural sector.[21] Thus the coverage of goods sold and purchased by the agricultural sector is not only more exhaustive but also includes those items which were not considered in the earlier studies.

Estimation of Weights

To find the weights for different items traded between the agricultural and the non-agricultural sectors, estimates of per capita consumption expenditure (NSS), final (private) consumption of the CSO and estimates of gross capital expenditure available in the Debt and Investment Survey of the RBI and those constructed by the CSO were used. Recognising that NSS (blown-up and CSO estimates differed in respect of many items quite markedly, weights of different items were worked out on two different bases for the construction of price indices. In one, estimates of private final consumption expenditure were those based on NSS consumption expenditure; in the other, estimates of final consumption expenditure by the agricultural sector were worked out on the basis of CSO estimates of private final consumption.

From the NSS estimates of per capita consumption for 1970-71 for rural and urban areas, first estimate of per capita consumption for the non-agricultural sector was worked out. Recognising that NSS estimates were at the level of retail prices, an adjustment was made in respect of each item to derive the estimates at producer prices.[22] The share of each of six commodity groups sold by the agricultural sector to the non-agricultural sector for final consumption was then worked out. For the individual commodity under the group cereal and cereal substitutes, share of each commodity was estimated from the details available from the 25th round of NSS.

In case of the principal raw materials supplied by agricultural sector to the non-agricultural sector for intermediate use, average of the value of the sales for the triennium ending 1971-72 was used for finding the share of each item in the total sales for intermediate consumption.[23] To combine the two series (final and intermediate consumption), weights used were the share of each group of items in the total sales to non-agricultural sector during the triennium ending 1971-72.

Per capita consumption estimates of different items are available for rural and urban areas from 25th round of NSS (1970-71). Consumption pattern of rural areas was assumed to hold good for the agricultural sector. The share of each commodity group identified as purchased by the agricultural sector was worked out in the total. However, before estimating the shares, estimates for items (a) edible oils (b) sugar and gur (c) pan tobacco and (d) fuel, power and light group was adjusted for the home grown component on the basis of NSS data.[24] From the data available in NSS report No 289 for the group fuel and light, the share of each item covered under it—kerosene, electricity, coal and matches—was worked out. By multiplying the percentage shares of each item in the group by the percentage share of the group as a whole in the total expenditure, the share of each individual item was estimated.

Similarly, in case of the group miscellaneous goods and services, the share of items covered in the group, namely, (a) drugs and medicine, (b) cosmetics and toilet articles, and (c) consumer services, were estimated. In case of durable goods, metal products, utensils and bicycles were taken as items representing this group and it was assumed that each of these four items had equal weights.

Estimates of value of inputs used in the agricultural sector for production are available for the National Accounts Statistics (CSO 1975, Disaggregated Tables), Separate estimates for salt and oil cakes, are however, not available in CSO publication; these are included in the group feed for livestock. However, using the same methodology as adopted by the CSO the estimates of total purchases of these two items were worked out for 1969-70, 1970-71 and 1971-72.[25] From the average expenditure on the 7 items purchased by the agricultural sector for intermediate use during the period 1969-70 to 1971-72, the weight of each item in the total purchases was worked out.

In case of items purchased for capital formation by the agricultural sector, recognising that for certain items purchases were made only for some components and not for all their proportions were worked out under certain assumption.[26] Again, for items involving construction, it was assumed that demand for different purchased items would be in the same proportion as in the construction of buildings. On this basis, weights for each purchased item were worked out. Again, to find the weight of the purchases made by agricultural sector for capital formation in the total purchases (final, intermediate and capital formation) made by this sector, the ratio of purchased items to gross total capital formation was estimated from the data available in Debt and Investment Survey. The estimates of purchases by agricultural sector for capital formation were obtained by multiplying this ratio with the CSO estimates. This was divided by the estimate of total purchases (final consumption + intermediate use + capital formation) to obtain weight of the purchases of this group in the total. Again, to derive weights for the purpose of combining price indices of different commodity groups on the basis of CSO estimates of private final consumption expenditure, which are not available separately for agricultural and non-agricultural sectors; estimates of private consumption for the two sectors were arrived at by multiplying the CSO estimates with the share of agricultural and non-agricultural sectors in the consumption of each item estimated from NSS data. These estimates of private final consumption were then used for working out the weights.

Again, as NSS data on consumption expenditure was not available for all the years for calculating weights for each year, CSO estimates of final consumption expenditure were used for this purpose. Following the above procedures, estimates of purchases and sales by the agricultural sector of different commodity groups for each year from 1967-68 to 1976-77 and the share of each commodity group in total purchases/sales was calculated.[27] These weights were used for constructing price indices using Paasche's formula.

Price Indicators

Realising the limitations of index number of wholesale prices in reflecting truly the movement of prices received by the farmers for commodities sold to the non-agricultural sector, these were not used as price indicators. Farm harvest prices were considered as indicators of prices received in the case of cereals, oilseeds, fibres, gur and tobacco.[28] All-India farm harvest price for each commodity was worked out as weighted average price of state farm harvest prices—weights being the share of each state in the total all-India production of that commodity.[29] As farm harvest prices for the commodities/commodity groups as rubber, milk and milk products, vegetables and fruits, meat, fish and egg and gram and pulses were not available, implicit price indices were worked out by dividing the value of output (CSO estimates) at current prices by those at constant prices. In case of sugarcane, all-India price was estimated for each year as weighted average price based on the prices paid by factories in different states, using share of the state in the total cane crushed by factories in the country as a weight.

Although doubts have been raised regarding the capacity of wholesale prices to reflect truly the changes in the price paid by consumers[30] even in the case of goods of industrial origin, in the absence of any other price series on a uniform and representative basis, we had to resort to the use of wholesale price index numbers for commodities purchased by the agricultural sector.[31] This would impose some limitation on the validity of our results in as much as they would discriminate in favour of non-agricultural sector.[32]

Choice of Base Year And Method of Constructing Price Indices

Agricultural production and, consequently, the prices of agricultural commodities fluctuate more widely than those of the non-agricultural commodities. Therefore, a three-year base period would provide a better base for comparison than a single-year base. Since agricultural production both in case of foodgrains and commercial crops remained fairly stable during 1969-70 to 1971-72 (Table 8.3) agricultural prices did not fluctuate very widely in case of large number of crop.[33] Recognising that the farm prices in this period of three years would provide a more stable and a reasonable base for comparison, it was decided to treat the prices of agricultural and non-agricultural commodities for the triennium ending 1971-72 (financial year) at par.[34]

TABLE 8.3 : All-India Index Numbers of Agricultural Production

(Triennium ending 1969-70=100)

Year	*Foodgrains*	*Non-Foodgrains*	*All Crops*
1	2	3	4
1967-68	98.7	99.0	98.9
1968-69	97.3	97.4	97.3
1969-70	104.0	103.6	103.8
1970-71	112.9	108.7	111.5
1971-72	111.4	110.9	111.2
1972-73	102.5	102.5	102.3
1973-74	110.3	117.0	112.4

Both Laspeyre's and Paasche's formulae were used for constructing indices of prices received and paid. The estimated indices of prices received and paid by the agricultural sector and of the net barter terms of trade based on Laspeyre's approach, with weights derived from NSS data on consumption expenditure for combined indices of different groups, are presented in Tables 8.4 to 8.6; those based on weights derived on the basis of CSO estimates of private final consumption expenditure are provided in Tables 8.7 to 8.9. Estimates of prices received, paid and of net barter terms of trade based on Paasche's approach are presented in Tables 8.10 to 8.12.

(a) Laspeyre's formula and weights derived from NSS data

It would be seen from Table 8.4 that the prices received by the agricultural sector for all commodities taken together showed an increases of about 63 per cent during the period 1967-68 to 1977-78. The prices of commodities sold for intermediate consumption rose at a rate faster

TABLE 8.4 : Indices of Prices Received by the Agricultural Sector

(Triennium ending 1971-72=100)

	Prices Received for						
Year	*Commodities Sold for Final Consumption*				*Commodities Sold for Intermediate Conusmption*[4]	*All Field Crops*[5]	*All Commodities**
	Cereals[1]	*Field Crops*[2]	*Subsidiaries*[3]	*All*			
1	2	3	4	5	6	7	8
1967-68	112.5	113.3	88.2	107.4	89.4	102.10	100.15
1968-69	100.8	98.4	92.7	97.0	86.4	92.78	92.73
1969.70	97.9	99.5	96.0	98.7	92.2	96.08	96.08
1970-71	100.6	98.2	100.8	98.8	102.9	100.40	100.46
1971-72	101.5	102.5	103.6	102.7	102.1	102.31	102.46
1972-73	112.5	116.4	113.3	115.7	118.2	117.24	116.71
1973-74	141.3	144.3	135.8	142.3	143.6	143.97	142.82
1974-75	190.3	183.0	158.9	177.3	149.1	167.11	165.94
1975-76	155.5	151.6	165.3	154.8	124.6	138.95	142.63
1976-77	144.9	145.0	160.0	148.5	163.3	153.58	154.46
1977-78	155.6	160.3	167.4	161.9	164.7	162.36	163.03

*Weighted average on NSS data for final consumption.

Note : Here and in subsequent tables.

1. Cereals covers rice, wheat, barley, maize, bajra, jowar and ragi.
2. Field crops cover cereals, pulse, vegetables and fruits.
3. Subsidiaries include milk and milk products, meat, fish and egg;
4. Intermediatary goods include cotton, jute, oilseeds, sugarcane rubber and tobacco; and
5. All commodities consist of field crops+commodities sold for intermediate consumption.

than those sold for final consumption during this period. Within the group of commodities sold for final consumption, the prices of subsidiary products (milk and milk products, meat, fish and egg) had increased at a rate faster than those of cereals, pulses, vegetables and fruits taken together.

The indices of prices paid by the agricultural sector for commodities purchased for different purposes (Table 8.5) would show that prices of commodities purchased for capital formation experienced the sharpest rise during the period 1967-68 to 1977-78. The prices of goods purchased for final consumption increased by 108 per cent; and since this group had a lion's share in the total purchases of the agricultural sector, the indices of prices paid for all commodities showed a rise of 109 per cent during the same period.

The net barter terms of trade during this period moved against the agricultural sector owning to a sharper increase in the prices paid by this sector. It would be seen from Table 8.6 that since 1974-75 the terms of trade have moved against the agricultural sector. The sharpest deterioration in the terms of trade occurred in the sub-sector that consists of commodities sold for final consumptions. Whereas for the agricultural sector, inclusive of subsidiaries, the terms of trade moved from 115.6 in 1967-68 to 89.7 in 1977-78, for the main sector that consists of field crops only, the terms of trade moved from 117.8 in 1967-68 to 89.4 in 1977-78 which implies that deterioration in the terms of trade was more pronounced when subsidiaries were excluded from this sector. A comparison of these results with those derived from the calculations of Thamarajakshi (1977) by adopting the triennium ending 1971-72=100 showed that whereas for the period 1967-68 to 1974-75 (the common period between ours and her study) according to her data the terms of trade moved in favour of agriculture from 100.63 to 107.79, using the refined methodology, our results showed that during this period the terms of trade moved against agriculture from 117.8 to 101.6.

TABLE 8.5 : Indices of Prices Paid by the Agricultural Sector

(*Triennium ending 1971-72=100*)

Year	*Prices Paid for Commodities Purchased for*			
	Intermediate Consumption	*Capital Formation*	*Final Consumption*	*All* Uses*
1	2	3	4	5
1967-68	93.7	83.0	85.3	86.7
1968-69	93.8	85.2	87.1	88.2
1969-70	98.2	92.0	93.7	94.4
1970-71	100.3	100.2	100.6	100.5
1971-72	101.6	107.7	105.7	105.1
1972-73	109.6	117.2	113.0	112.7
1973-74	126.3	134.3	133.0	131.9
1974-75	188.1	159.5	161.9	166.6
1975-76	193.1	188.9	159.9	168.6
1976-77	191.5	194.3	165.7	173.0
1977-78	193.6	191.7	177.4	181.7

*Weighted indices based on NSS estimates of consumption expenditure.

TABLE 8.6 : Indices of Net Barter Terms of Trade

(Triennium ending 1971-72 = 100)

Years	*Between Prices Received and Paid for All Purposes*				*Between Pirces Received and Paid for Intermediate Consumption Goods*	
	Prices Received for Commodities Sold for				*Prices Received for*	
	Final Consumption	*Intermediate Consumption*	*Field Crops All Uses*	*All Agricultural Commodities*	*Field Crops*	*All Agricultural Commodities*
1	2	3	4	5	6	7
1967-68	123.9	103.1	117.8	115.6	109.0	106.9
1968-69	110.0	98.0	105.2	105.1	98.9	98.9
1969-70	104.6	97.7	101.8	101.8	97.8	97.8
1970-71	98.3	102.4	99.9	100.0	100.1	100.2
1971-72	97.7	97.2	97.3	97.5	100.7	100.9
1972-73	102.7	104.9	104.0	103.6	107.0	106.5
1973-74	107.9	108.9	109.2	108.3	114.0	113.1
1974-75	106.4	89.5	100.3	99.6	88.8	88.2
1975-76	91.8	73.9	82.4	84.6	72.0	73.9
1976-77	85.8	94.4	88.8	89.3	80.2	80.7
1977-78	89.1	90.6	89.4	89.7	83.9	84.2

*Weighted indices based on NSS estimates of consumption expenditure.

TABLE 8.7 : Indices of Prices Received by the Agricultural Sector

(Triennium ending 1971-72 = 100)

Years	*Sale for Final Consumption*				*Intermediate Consumption*	*All Field Crops*	*All Commodities**
	Cereals	*Field Crops*	*Subsidiaries*	*All*			
1	2	3	4	5	6	7	8
1967-68	112.5	115.2	88.2	108.9	89.4	102.4	101.9
1968-69	100.6	98.6	92.7	97.2	86.4	93.4	93.3
1969-70	97.9	99.0	96.0	98.3	92.2	96.1	96.1
1970-71	100.6	97.9	100.8	98.5	102.9	100.0	100.1
1971-72	101.5	103.2	103.6	103.3	102.1	102.7	102.9
1972-73	112.5	117.4	113.3	116.4	118.2	117.7	117.1
1973-74	141.3	144.5	135.8	142.5	143.6	144.1	142.9
1974-75	190.3	182.3	158.9	176.8	149.1	168.2	166.8
1975-76	155.5	151.5	165.4	154.8	124.6	140.1	143.9
1976-77	144.9	145.4	160.1	148.8	163.3	153.0	154.0
1977-78	155.6	159.8	167.6	161.6	164.7	161.9	162.7

*Weighted average on CSD basis.

(b) Laspeyre's formula and weights derived from CSO data

Using Laspeyre's approach and the weights derived from the CSO data, the estimates of prices received, prices paid net barter terms of trade are presented in Tables 8.7 to 8.9. It would be seen from these tables that the difference between these estimates and those based on weights derived from NSS data alone was only marginal in most cases. However, even on the basis of these estimates, the net barter terms of trade have moved against agricultural sector since 1975-76; the deterioration in terms of trade was particularly more pronounced against the main sector consisting of field crops. It was noted that in the 1970s the terms of trade moved in favour of agriculture only in 1972-73 and 1973-74 when agricultural prices ruled high due to short fall in production.

TABLE 8.8 : Prices Paid by the Agricultural Sector

(Triennium ending 1971-72=100)

Year	*Commodities Purchased for*			
	Final Consumption	*Intermediate Consumption*	*Capital Formation*	*All Uses*
1	2	3	4	5
1967-68	85.7	93.7	83.0	86.5
1968-69	87.6	93.8	85.2	88.2
1969-70	93.8	98.2	92.0	94.2
1970-71	100.5	100.3	100.2	100.5
1971-72	105.7	101.6	107.7	105.3
1972-73	113.3	109.6	117.2	113.1
1973-74	132.2	126.3	134.3	131.6
1974-75	161.0	188.1	159.5	164.2
1975-76	160.3	193.1	188.9	165.9
1976-77	165.5	191.5	194.3	170.3
1977-78	176.1	193.6	191.7	180.1

Note : Weighted average (CSO basis) of the price indices.

(c) Paasche's formula and weight based on CSO data

The results obtained by using current year weights, which have the advantage of comparing what has happened to the unit values with the base period and reflect current quantities are given in Tables 8.10 to

8.12. Here, it needs to be mentioned that a comparison of any two years which does not have common based period refers to different quantities of trade and involves compositional as well as unit value change.

TABLE 8.9 : Indices of Net Barter Terms of Trade

(Triennium ending 1971-72=100)

Years	*Between Prices Received and Paid for All Purposes*				*Between Prices Received and Paid for Intermediate Consumption Goods*	
	Prices Received for Commodities Sold for				*Prices Received for*	
	Final Consump-tion	*Intermediate Consumption*	*Field Crops All Uses*	*All Agricultural Commodities*	*Field Crops*	*All Agricultural Commodities*
1	2	3	4	5	6	7
1967-68	125.9	103.4	120.5	117.8	111.2	108.8
1968-69	110.2	98.0	105.9	105.8	99.6	99.5
1969-70	104.4	97.9	102.0	102.0	97.9	97.9
1970-71	98.0	102.4	99.5	99.6	99.7	99.8
1971-72	98.1	97.0	97.5	97.7	101.1	101.3
1972-73	102.9	104.5	104.1	103.5	107.4	106.8
1973-74	108.3	109.1	109.5	108.6	114.1	113.1
1974-75	107.7	90.8	102.4	101.6	89.4	88.7
1975-76	93.3	75.1	84.5	86.7	72.6	74.5
1976-77	87.4	95.9	89.8	90.4	79.9	80.4
1977-78	89.7	91.6	89.9	90.3	83.6	84.0

Note : Weighted indices based on CSO estimates of consumption expenditure.

TABLE 8.10 : Indices of Prices Received by the Agricultural Sector

(Triennium ending 1971-72=100)

Year	*Final Consumption*	*Intermediate Consumption*	*All Commodities*
1	2	3	4
1970-71	98.4	103.8	101.8
1971-72	103.5	102.2	101.6
1972-73	117.4	119.0	114.2
1973-74	142.6	149.9	145.1
1974-75	175.5	154.0	176.3
1975-76	155.2	127.4	147.0
1976-77	149.6	167.8	152.3

Note : Based on current year weights.

TABLE 8.11 : Indices of Prices Paid by Agricultural Sector

(Triennium ending 1971-72=100)

Year	*Prices Paid for Commodities Purchased for*			
	Intermediate Consumption	*Capital Formation*	*Final Consumption*	*All Uses*
1	2	3	4	5
1970-71	100.3	100.2	100.6	100.5
1971-72	101.4	107.7	106.1	105.8
1972-73	109.6	117.2	113.7	113.3
1973-74	129.9	134.3	132.7	132.3
1974-75	192.7	159.5	161.4	166.9
1975-76	199.9	188.9	106.2	168.5
1976-77	195.7	194.3	164.4	171.8

Note : Based on current year weights.

It would be seen from Table 8.10 that the index of prices received by the agricultural sector moved from 101.8 in 1970-71 and 152.3 by 1976-77 whereas in case of fixed base index (Table 8.7) it moved from 100.10 in 1970-71 to 154.0 in 1976-77. In contrast, for the years 1973-74, 1974-75 and 1975-76 when inflation was high, the current weight indices showed a larger order of increase as compared with the fixed base indices.

TABLE 8.12 : Indices of Net Barter Terms of Trade

(Triennium ending 1971-72=100)

	Between Prices Received and Paid for All Purposes			*Between Pirces Received and Paid for Intermediate Commodities*
	Prices Received for Commodities Sold for			
	Final Consumption	*Intermediate Consumption*	*All Commodities*	*Prices Received for All Agricultural Commodities*
1	2	3	4	5
1970-71	97.9	103.8	101.3	101.5
1971-72	98.0	96.8	96.2	100.2
1972-73	103.6	105.0	108.8	104.2
1973-74	107.8	112.6	109.7	112.3
1974-75	105.2	92.3	105.6	91.5
1975-76	92.1	75.6	87.2	73.7
1976-77	87.1	97.7	88.7	77.8

Note : Based on current year weights.

The prices paid by the agricultural sector for goods purchased for intermediate use are given in Table 8.11 which shows that price index moved from 100.3 in 1970-71 to 192.7 to 1974-75, 199.9 in 1975-76 and 195.7 in 1976-77 whereas the indices derived using Laspeyre's formula stood at 100.3 in 1970-71 and at 188.1, 193.1, 191.5 in 1974-75, 1975-76 and 1976-77 respectively. This supports the earlier hypothesis that the adoption of Laspeyre's formula (base year weights) would underestimate the rise in the prices paid by the agricultural sector. The indices of net barter terms of trade based on Paasche's formula (current year weights) in Table 8.12 show that the net barter terms of trade moved against agricultural sector after 1974-75 and for 1975-76 and 1976-77 the index of net barter terms of trade stood at 87.2 and 88.7 respectively.[35] A comparison of these indices with those calculated using Laspeyer's formula in Table 8.9 indicates that Laspeyer's formula inflated indices of net barter terms of trade in the years when they were in favour of agriculture but underestimated the indices for the years when the terms of trade went against the agricultural sector. In fact, the difference between the indices based on Laspeyre's and Paasche's is about 2 percentage points just after 5 years or so indicating that this underestimation of adverse movement of terms of trade could be of a much larger order in the series which involved comparison over 15-20 years.

Estimated Inter-Sectoral Terms of Trade

Adopting the improved methodology discussed in this paper, inter-sectoral terms of trade were worked out which are updated every year by the Commission for Agricultural Costs and Prices.

To summarise this section, the agriculture sector purchases commodities from the non-agriculture sector broadly for three purposes:

(1) Consumption, (2) Production and (3) Investment. From the data available from the National Sample Survey, Central Statistical Organisation and Reserve Bank of India reports, 33 items purchased by the agricultural sector were identified and their relative weights estimated.

Looking at the other side of the coin, the agricultural sector broadly sells to the non-agricultural sector such commodities as wheat, jowar, bajra, maize, ragi, paddy, rice, gram/pulses products, milk and milk-products, meat and meat products, fruits and vegetables, gur, groundnut, rapeseed and mustard, linseed, sesamum, castor seed, jute, cotton, sugarcane, tobacco and rubber. Considering the weights assigned to these commodities in the construction of Economic Advisor's Index number of wholesale prices, the relative shares of 21 commodities were worked out for constructing the new index of prices received by the farmers. For estimating the terms of trade between agricultural sector and non-agricultural sector, not only a list of commodities exchanged between the two sectors is required, but also the relative share of each commodity in the total trade.

The estimated terms of trade in Table 8.13 show that terms of trade have been unfavourable to agricultural sector even during the eighties.

TABLE 8.13 : Indices of Prices Received Prices Paid and Terms of Trade, Agriculture Sector, 1961-62 to 1991-92

Year	*Prices Received*	*Prices Paid*	*Terms of Trade R/PX 100*
1961-62	51.5	63.9	80.6
1962-63	52.8	66.3	79.6
1963-64	56.6	77.6	72.9
1964-65	67.2	71.5	94.0
1965-66	77.7	75.5	102.9
1966-67	93.7	83.0	112.9
1967-68	100.2	86.7	115.6
1968-69	92.7	88.2	105.1
1969-70	96.1	94.4	101.8
1970-71	100.5	100.5	100.0
1971-72	102.5	105.1	97.5
1972-73	116.9	112.9	103.5
1973-74	145.0	132.3	109.6
1974-75	166.8	166.9	99.9
1975-76	142.4	168.3	84.6
1976-77	157.0	173.2	90.6
1977-78	164.8	181.6	90.7
1978-79	157.1	183.9	85.4
1979-80	185.4	209.3	88.6
1980-81	213.6	244.8	87.3
1981-82	224.2	270.5	82.9
1982-83	237.2	279.9	84.7
1983-84	261.1	302.6	86.3
1984-85	276.8	321.8	86.0
1985-86	278.0	337.4	82.4
1986-87	308.0	361.2	85.3
1987-88	348.2	400.5	87.0
1988-89	359.7	422.2	85.2
1989-90	386.8	455.1	85.0
1990-91	439.3	505.0	87.0
1991-92	516.3	567.3	91.0

In this context, the Central Statistical Organisation data show that while the implicit index of prices at which gross domestic product is computed rose by 76 per cent between 1973-74 and 1984-85, the implicit price index of gross capital formation in agriculture rose by 169 per cent during the same period. Again, according to Commission for Agricultural Costs and Prices, between 1983-84 and 1986-87, the prices paid for capital formation increased over 19 per cent as against an increase of 13 per cent in the prices received by the farm sector. It is indeed a matter of serious concern that capital formation in the agricultural sector has become much more costly. The price that the farmer has to pay in making capital investments for further improvement of his farm organisation is highly important for sustained future growth in productivity and farm income.

Policy Implications

Owing to non-availability of retail prices of the commodities purchased by the agricultural sector, index number of wholesale prices, which has its own limitations, has been used in this study. To remedy this situation, information on prices paid by the agricultural sector for different commodities at retail level needs to be collected. Again, to derive weights of different items purchased by the agricultural sector with greater precision, exhaustive expenditure survey of cultivating households need to be carried out. There is thus an urgent need for developing a project which provides the requisite data for the construction of indices of prices received and prices paid by farmers.

The effect of terms of trade turning against the agricultural sector on the development of agriculture may not be significant in a subsistence agriculture where the proportion of purchased inputs by the agricultural sector is very low. However, with the proportion of purchased inputs rising with the modernisation of agriculture, the adverse terms of trade can substantially affect the level of technological inputs used and hence capital formation and level of output. Such a development would be counter productive in the early stages of the modernisation of agriculture. It is, therefore, important that the movement of terms of trade are kept under control in the interest of rapid development of agriculture.

It needs to be appreciated that the parity can fluctuate sharply in the short period, but the price parity cannot be adjusted to such fluctuations which may tend to cancel out over a period. A correction factor for adverse terms of trade can be considered only after a medium terms trend was established.

Note : Reproduced from *Economic and Political Weekly:* Review of Agriculture, December 1980, pp. A-173-A 184.

NOTES

1. See, for example, Dandekar (1968), Johl (1974), Hopper (1976) and Tyagi (1979).
2. Kindleberger (1956) defines (in the context of international trade) the capacity to import as the net barter terms of trade times the quantity exported.
3. For a detailed discussion of the problem of constructing and interpreting index number of unit values, see Allen (1953).
4. Though fuller version of the Venkataraman's (1979) study is not available, Rao (1979) quoting the results of this study mentioned that the basic methodology used was the same as that adopted by Thamarajakshi (1969).
5. It is questionable to regard the entire range of basic industrial chemicals as commodities purchased by the agricultural sector. In fact, Sethi (1971) admits that the manufacturers price index with which the agricultural price index is compared includes a number of industrial commodities which farmers do not directly consume.
6. Thamarajakshi's study appears to be based on this limited coverage only.
7. Although Thamarajakshi (1969) made adjustments for netting the final consumption estimates for imports into the country of such items as foodgrains, no mention for an adjustment required on this account is made.
8. For example, in case such commodities as vegetables, fruits and milk where the price spread was higher than, say in cereals, the weights for the former wou!d turn out to be over-estimates and those for the latter underestimates.
9. Limitations to the use of the index number of wholesale prices arise mainly from the following factors : (i) equal weight attached to all the centres selected for getting price quotations; (ii) use of issue prices as price quotations for certain centres; (iii) use of single year as base year, and (iv) equal weight attached to all the weeks.
10. Thamarajakshi (1969).
11. Total exports from the agricultural sector were estimated by adjusting the sales for final consumption for imports of these commodities into the country and then adding to it estimated value of sales of commodities for intermediate consumption.
12. Estimates of final consumption expenditure on agricultural and non-agri- cultural commodities have been worked out following Thamarajakshi's approach.
13. Using Thamarajakshi's data but considering the adjusted value of R_T it was

found that while the uncorrected R_T gave a growth rate of 1.69 per cent per annum for the period 1954-55 to 1960-61, for the value of exports it was only of the order of 0.88 per cent per annum when adjusted R_T was used.

14. To illustrate how this could happen hypothetical indices of term of trade and income terms of trade are presented in Table A.

TABLE A : Hypothetical Indices

Year	*Agricutlural Sector*			*Non-Agricultural Sector*		
	Net barter -terms of trade	*Indices of export*	*Income terms of trade*	*Trade*	*Export*	*Income terms of trade*
1.	98	99	97	102	97	98.9
2.	100	99	99	100	98	98.9
3.	99	99	98	101	99	100
4.	100	100	100	100	100	100
5.	101	100	101	99	102	101
6.	103	101	104	97	106	103
7.	105	102	107	95	110	105

The table indicates that income term of trade of both the sectors have improved.

15. During the period 1960 61 to 1970-71 whereas the value of all inputs used in the agricultural sector increased by about 129 per cent that of fertilisers escalated by 624 per cent and during this period the fertiliser consumption showed a three-fold increase. Similarly, whereas in 1961-62 the share of wheat in the per capita consumption of cereals in Urban areas was 32.8 per cent, it increased to 36.3 per cent by 1970-71. In 1961 the share of imported wheat in the total availability was 24.1 per cent whereas in1971 total consumption requirements were met out of the domestic production.

16. During the period 1970-71 to 1977-78 the prices of fertilisers rose by 78 per cent and during this period its application per hectare of net cropped area showed an increase of 83 per cent. Its share in the total inputs increased from 38.3 in 1970-71 to 42.2 by 1976-77.

17. It would be seen from Table B that prices increased sharply when the production of groundnut, cane, cotton, rice declined.

TABLE B

Crop	*Year*	*Production Index*	*Price Index*
Sugarcane	1970-71	107.1	100.00 (gur)
	1971-72	97.9	152.4 (gur)
Cotton	1975-76	94.5	136
	1976-77	88.5	198
Groundnut	1971-72	103.5	88
	1972-73	96.3	104
Rice	1971-72	102.0	103
	1972-73	99.2	116

18. Incidentally, the list of repeated users of this result includes Dantwala (1976) who himself has pondered over the issue of base and terminal years in the calculation of compound growth rates, at great length.
19. The commodities/commodity groups identified were edible oils, sugar, salt, kerosene, matches, electricity, coal, tobacco and tabacco products, textiles, footwear drugs and medicine, cosmetics/soap and detergent, metal products, paper and paper products, utensils, cycles and consumer services.
20. The details of the items required in construction are available in "All India Rural Debt and Investment Survey : Tangible wealth. Capital expenditure and capital formation of Rural Households". *RBI Bulletin,* June 1965. The items identified were cement, lime, transport equipment, machinery and machine tools, iron and steel, log and timber, tools and implements, bricks and tiles.
21. The 22 commodities were rice, wheat, jowar, bajra, barley, maize, ragi, milk and milk products, meat/fish and egg, vegetables, and fruits, gram and pulse products, gur, groundnut, sesamum, rapeseed and mustard linseed, castor seed, cotton, jute and mesta, rubber, sugarcane and tobacco.
22. The results of empirical studies on the subject of price spread show large variations and estimates of price spread for all commodities for the same year are not readily available. In absence of uniform data the difference between the producer and retail prices was assumed to be 15 per cent in case of cereals and gur; 30 per cent in case of pulses and pulse products and 40 per cent in case in meat fish and eggs and vegetable, fruits etc.
23. In case of oilseeds first estimate of production and net of seed requirements were worked out by deducting the seed requirement as the acreage under the crop in the subsequent year into the seed rate. This net production estimate was then adjusted for consumption in raw (i.e., in the seed form) in rural areas. This net estimate was then evaluated at the average farm harvest price for finding the value of sales. In case of edible oils, the total value was then adjusted for consumption in the form of oil (home produced) taking the ratio of home produced oil to total oil consumption in rural area as 1.23/1.26. In case of fibres, cotton, jute and mesta, estimates of extra factory consumption are available. In case of cotton, 70 per cent of extra factory consumption was assumed to have been consumed in the agricultural sector.
24. On the basis of NSS data (Report 239) in case of oilseeds, 2.4 per cent was taken as home grown while in case of sugar and gur it was 56.5 per cent for fuel and power 76.7 per cent and 3.2 per cent for pan tobacco.
25. Expenditure on salt and medicine is assumed at one rupee per animal; 55 per cent of groundnut oil cakes production, 80 per cent of linseed oil cakes production and 60 per cent of coconut and cotton seed oil cakes production are taken to have been purchased by the agricultural sector for animal feed. See National Accounts Statistics (CSO 1961).
26. It was assumed that in case of wells, purchases would be of the order of 50

per cent whereas for other irrigation resources, farm houses and barns, etc, and residential plots and buildings this ratio would be 67, 60 and 66 per cent respectively.

27. Estimates of final consumption were worked out by dividing the CSO estimates of final consumption expenditures for each commodity in the same ratio as observed in the NSS data (1971) for agricultural and non-agricultural sectors. These then were adjusted for consumption out of imports. The estimates of sales commodity-wise by the agricultural sector for intermediate use were worked out in the same way as for the period 1969-70 to 1971-72. In case of goods purchased for capital formation, the estimates of total purchases were worked out from the CSO estimates of gross capital formation, assuming that the share of purchased item has remained same as in 1971-72 since for any latter period details of the capital formation item-wise were not available.
28. Farm harvest price of a commodity is defined as the average wholesale price at which commodity is disposed of by the producer to the trader at village site during the specified marketing period after the commencement of harvest. See Directorate of Economics and Statistics (1975).
29. Ideally, the share of each state in the all-India marketed surplus of that commodity should have been used as weight. Since state-wise estimates of marketed surplus are not available, production shares have been taken as proxy for shares in the marketed surplus.
30. On the inadequacy of index numbers of wholesale prices in reflecting movement of prices of manufacturers, see Tyagi (1979); and for their inadequacy in reflecting changes in the consumer prices see *EPW, (1973)* and *Indian Labour Journal* (1977).
31. Realising that footwear purchased by the agricultural sector are not those manufactured in the organised industrial sector, in place of Economic Adviser's indices for footwear, which are based on quotation from Bata and Flex Shoe Companies, indices for tanned cured and finished leather were used as proxy for footwear. Since Economic Adviser's index number are not available for consumer services, repair and maintenance and for educational expenses it was decided to use all-commodities index numbers for the former and of paper and paper products for the latter as price indicators.
32. As retail prices are expected to increase at a rate faster than that by which wholesale prices rise due to fast increasing transportation costs and trade margins, the use of wholesale price indices in case of manufactures may show a rise of lower order than that by which the prices paid by the farmers in fact have risen.
33. During the years 1969-70, 1970-71 and 1971-72 the index numbers of wholesale prices (Base 1961-62=100) for agricultural commodities remained at 201, 195 and 198 respectively.
34. This base period by and large satisfies the criteria spelt out by National Commission on Agriculture (1976).

35. Based on Fisher's index formula, indices of net barter terms of trade stood at 87.0 for 1975-76 and at 89.5 for 1976-77 as Against 100.5 for the year 1970-71. These estimates are not very different from those obtained by using Laspeyre's and Passche's formulae.

REFERENCES

Allen, R. G. D. (1953), 'Index Numbers of Volume and Prices' in *International Trade Statistics*, (eds.) R. G. D.. Allen and J. Edward Ely, Wiley, New York.

CSO (1961), "National Income Statistics : Proposals for a Revised Series of National Income Estimates for 1955-56 to 1959-60", Government of India, New Delhi.

CSO (1975), National Accounts Statistics 1960-61 — 1972-73 Disaggregated Tables", Government of India, New Delhi.

CSO (1978), "National Accounts Statistics", Government of India, New Delhi.

Chaudhary, Uma Roy (1967), 'Study of Trends in Consumption Expenditure 1953-54 to 1960-61', *Indian Economic Review.*

Chaudhary, Uma Roy, Raman, M. S. and Rao A. R. (1979), 'Trends in Private Consumption Expenditure from 1950-51', Paper presented at the 18th All-India Annual Econometric Conference, Srinagar.

Dandekar, V. M. (1968), 'Agricultural Price Policy : A Critique of Dantwala', *EPW*, March 16.

Dantwala, M. L. (1967), 'Incentives and Disincentives in Indian Agriculture,' *Indian Journal of Agricultural Economics,* Vol. XXII, No. 2.

Dantwala, M. L. (1976), 'Agricultural Policy Since Independence', *Indian Journal of Agricultural Economics.* Vol. XXXI, No 4.

Dar, Ashok (1968), 'Domestic Terms of Trade and Economic Development of India 1952-53 to 1964-65', *Cornell International Agricultural Development Bulletin 12*, Ithaca, New York.

Directorate of Economics and Statistics (1975), Farm (Harvest) Prices of Principal Crops in India (1970-71 to 1974-75)'. Directorate of Economics and Statistics, Ministry of Agriculture and Irrigation, New Delhi.

EPW, (1973), "Lying Indices", *EPW* May 26.

Griffin, Keith (1974), *The Political Economy of Agrarian Change*, MacMillan.

Hopper, W. David (1976), 'Sixth Coromandel Lecture: A perspective on India's Food Production', *Eastern Economist.* January 30.

Johl, S. S. (1974), 'Foodgrain Prices Policy in India', Department of Economics and Sociology. Punjab Agriculture University, Ludhiana.

Kindlelberger, Charles P. (1956), *The Terms of Trade — A European Case Study*, John Wiley, New York.

Labour Journal (1977), *Indian Labour Journal,* September 1977.

Lipton, Michel (1977), "Why Poor People Stay Poor", Temple Smith, London.

Mason Edward (1966), *Economic Development of India and Pakistan,* Harvard University Press, Cambridge.

Mitra Ashok (1977), *Terms of Trade and Class Relations*, Frank Cass, London.

Mukherjee (1969), "National Income of India, Trend and Structure", Statistical Society of India, Calcutta.

NSS (1975) No. 269 : Tables with Notes on Consumer Expenditure, Ministry of Planning, Government of India.

NSS (1978) No. 239 : Consumer Expenditure for Cultivator Households: Rural (Special Requirement of the Ministry of Agriculture), Ministry of Planning, Government of India.

Rao, V. K. R. V. (1979), 'How Farmers have Ridden the Crest of Inflation' *Capital,* December 3.

RBI (1965) : All India Debt and Investment Survey (1961-62) Tangible Wealth, Capital Expenditure and Capital Formation of Rural Households RBI Bulletin June.

SECTION III

LABOUR ABSORPTION

9

TRENDS IN EMPLOYMENT IN INDIAN AGRICULTURE, LAND AND ASSET DISTRIBUTION

SHEILA BHALLA

It now appears that much of the technological change in agriculture in the past decade or so has been advantageous basically because it has been land-augmenting. Simultaneously, it has also been, on balance, labour-saving; so much so that despite cropping pattern shifts in favour of relatively more labour intensive crops in many States (together with marginal increases in the gross cropped area in most States), the combined effect of labour-saving technological change for individual crops plus the increases in area under relatively labour intensive crops has been that total labour absorption has either been stagnant or may even have fallen in absolute terms in several of the most technologically dynamic States. State level analysis suggests that these developments are associated with rapid mechanisation and increases in the use of chemical fertilisers and insecticides.[1] The net result is that despite the fact that three States (Andhra Pradesh, Gujarat and Maharashtra) continued to enjoy high and significant trend rates of growth in total labour absorption, the average elasticity of employment with respect to the production of 'all crops[2] for the period 1971-72 to 1983-84 is considerably below the corresponding[3] figure estimated by Tyagi (1981) for the ten-year period ended 1978-79: an employment elasticity of only 0.5945[4] for the later period compared to 0.769 for the earlier one.

The introduction of labour-saving technology in agriculture is a development which might well have been expected in the light of the historical experience of other countries, but in India it has come at a rather bad time. The shift of workers to productive non-agricultural occupations has only just got underway in proportionate terms, 'and has yet to result in a decrease in the absolute numbers engaged in agriculture. And in most states the introduction of labour-saving technology has begun even before the initial rising trend in labour absorption per hectare had time to exert effective upward pressure on the real wage rates of a burgeoning class of landless workers. For those who own little or no land and have an inadequate earnings base in terms of productive non-land assets, it is clear that agricultural growth by itself is increasingly unlikely to provide adequate incomes. To alleviate the poverty of the have-nots, direct measures may be necessary. As Krishnaswamy recently put it: "Many changes in production structure, institutions and policy will be required merely to ensure that growth and technological change do not cause, in the immediate future, more unemployment or more inequality."[5]

The main focus of this paper is on the employment aspect, because in my view, it is this which provides the context within which questions of equity in the distribution of land, and other productive assets in agriculture now have to be judged.

The section on labour absorption in Indian agriculture is, therefore, placed first. It has two parts. Part I raises broad questions of policy and interpretation in the larger historical and empirical setting. Part II deals with Indian evidence at the State level for the period 1971-72 to 1983-84. The estimates of labour absorption are based on data available from the Comprehensive Scheme for Studying the Cost of Cultivation of Principal Crops.

Questions of equity in the distribution of land and other assets are raised in Part III. The empirical base is National Sample Survey (NSS) data on ownership holdings, and, on assets, the series of decennial surveys of household indebtedness and investment initiated by the Reserve Bank of India (RBI) and subsequently conducted by the National Sample Survey Organisation (NSSO).

I. LABOUR ABSORPTION IN INDIAN AGRICULTURE

The Larger Historical and Empirical Setting

I presume a long-term economic dynamics of the sort described by Kaldor[6] and Kuznets[7] in which the share of agriculture in both national product and also in the workforce falls overtime; one in which the productivity of labour in agriculture is rising (partly because of yield improvements and the extension of gross cropped area (GCA) and partly because of the acceleration of the proportionate shift of workers to jobs outside of agriculture); and one in which the elasticity of agricultural employment with respect to total agricultural production declines and eventually becomes negative.

The facts in India do indeed coincide with the broad outlines of this description, except that in recent years the secondary sector, and manufacturing in particular may not be playing the job-creating role particularly effectively. Given a constant product wage in industry as a whole, Goldar's 1984 study[8] (covering the period 1960 to 1977) indicates that the short run elasticity of employment with respect to industrial output is only 0.254 although the long run elasticity is 0.799. However, industrial employment is extremely sensitive to variations in real labour costs.[9] Goldar finds that a one per cent increase in 'real labour costs' reduces employment by 0.344 per cent within one year, and over a long period, by more than one per cent. Largely because of a slow down in the rate of increase of 'real labour costs' in the seventies, labour absorption in industry increased in the second half of the period covered by the Goldar study. Preliminary work on data for more recent years suggests that employment elasticities with respect to industrial output may have declined subsequently in the late seventies and the early eighties.[10]

This is precisely the situation described by Ishikawa[11] in his discussion of the alternatives faced by the countries of South and South-East Asia. Given a low elasticity of demand for labour with respect to non-agricultural production; given scarce land and abundant labour within agriculture; and given that income and asset redistribution mechanisms do not work effectively, he argues that the only practical way of reducing poverty and improving rural income distribution may be to promote yield-increasing and land-augmenting technical change which is also labour-using.

Unfortunately for this policy prescription the prospects for inducing a shift in factor proportions in favour of labour in Indian agriculture now seem to be receding. It appears that the most effective yield-increasing and land-augmenting technologies currently available in India also happen to be labour-saving, at least in some regions and for some crop combinations. This being the case, the pursuit of the national objectives of increasing output and augmenting land resources comes into direct conflict with employment and equity goals. Moreover, in the Indian context it would be, in any case, virtually impossible to prevent cultivators from choosing the technological package which most increases their earnings.

What has happened in India is that for those crops and those States where the Green Revolution came early, the usual initial response was a sustained rise in labour use per hectare. This trend characteristically peaked in the mid-seventies or shortly afterwards, and subsequent increases in yield were associated with declines in per hectare labour absorption. However, in some States where improved practices and inputs were introduced, labour intensity continued to rise concurrently with yields and with total production.

On the face of it, the shape of some of these curves resembles the smooth graphs used by Ishikawa to illustrate the much longer period 'paths of change' in labour intensity in rice cultivation in Japan and Taiwan.[12] The rising phase of his inverted U-shaped curves is associated with a familiar set of labour-using, yield-enhancing technological factors: "irrigation and drainage, higher yielding varieties, and the application of larger amounts of fertilisers, and improved cultivation practices."[12] When the introduction of labour saving technological factors' (under which Ishikawa lists improved implements, and mechanisation), begins to outweigh the impact of labour-using factors, the labour absorption trend lines fall.

Ishikawa also argues that a 'late-comer' country is likely to adopt labour-saving technology much more quickly if that technology is already developed, as it may constitute the cheapest yield-increasing input combination available. The result is that the labour absorption graphs of the 'late-comers' are likely to peak at much lower levels of labour absorption. This possibility may need to be kept in mind when using Indian data on recent experience to project future trends.

Ishikawa's discussions, however, fail to focus explicitly on what seems to be the main force at work to induce labour-saving technological change in Indian agriculture during the past decade or so, partly because he sets much of his analysis in the context of comparative studies of labour absorption in the cultivation of a single crop-rice or of a rice-rice crop sequence.

Indian farmers, like their East and South-East Asian counterparts, adopt a labour-saving technology because it promises enhanced incomes. But it needs to be emphasised that chosen technologies increase earnings in relation to the entire *mix* of crops and operations and may not necessarily do so with respect to the particular crop whose employment elasticity with respect to yield is observed to go down or even to become negative (as has happened in the case of wheat cultivation in almost every major State in India) (see Table 4.2). Further, in most cases in India, the human labour-saving technology is also clearly *bullock* labour-saving; that is, crop land can be and is being released from fodder cultivation to more remunerative crops, and this may be the most widely operative factor in India today. But this factor appears normally to operate in combination with at least one additional land-augmenting practice in specific cases—about which more later.

At the same time, it needs to be noted that in the case of some of the 'early-comers' in India, the land-augmenting advantage of labour-saving technology may not have been its only attraction. Rising real wage rates, as in Haryana and Punjab, may have tilted the scales in favour of labour-saving technology in the early and the mid-seventies. But it is important to note that this cannot be generalised, even to Uttar Pradesh, where labour intensity curves for wheat, sugarcane (and possibly also rice) follow as clear an inverted U-shaped time trend as in Haryana and Punjab (when the latter's labour absorption curves are extended backward using data for the 1960). In the case of Uttar Pradesh, including West Uttar Pradesh in the hey-day of its wheat revolution, the constancy of real wages even at the regional level is well established.[14]

Today, in East Uttar Pradesh, labour-saving mechanisation is proceeding apace, again, *despite* constant real wages, and apparently abundant labour. In this case, aside from the possibility that Ishikawa's 'late-comer' effect is in operation, the other potential explanation needs to be mentioned. This is Haryana's[15] labour 'transactions costs' theory. He argues that, before Independence, the caste system worked as an efficient mechanism for supplying docile and reliable labour, which reduced the costs of hiring and managing labour in India. Since Independence, he argues, the caste system has instead increased labour conflicts and hence labour transaction costs. In this situation, "the sharp class conflicts among cultivators and agricultural labourers under the caste system," together with "increased labour demand from irrigation and new technology might induce mechanisation without creating much employment opportunities."[16]

While this possible explanation need not be rejected out of hand, in the presence of substantial gains from land-augmenting, labour-saving technology, it is clear that such technology should have been chosen even in the absence of the hypothesised 'transaction costs'. Moreover, there are much cheaper ways of reducing the costs of labour management, including piece rate systems, and a variety of forms of hiring through labour contractors. Consequently, in the search for explanations for the adoption of labour-displacing technologies in particular regions, the first step, in my view, should be to see whether or not its effect has been significantly land-augmenting, or yield-increasing. It may be that many instances of land-augmenting, labour-saving technological change have not been properly identified and documented.

In West Uttar Pradesh, for example, in November-December, the sugarcane *ratoon* crop is harvested. The practice, in pre-HYV wheat days was to let this land lie fallow until the *kharif* maize was sown. Now, as soon as the *ratoon* crop is cut, the farmers use a tractor-drawn disc plough, which cuts the tough *ratoon* roots into small pieces. This makes it possible for the cultivator to sow wheat immediately. With bullock-drawn conventional ploughs, it used to take two months to prepare the land for the next crop. Without mechanisation, it would not have been possible to grow the wheat crop at all. A parallel change in practice is said to have taken place in some cotton-wheat crop cycle areas of Punjab. In other cases it may simply be a question of timeliness which increases the yield of the second crop but is not land-augmenting, as in some of the wheat-rice crop cycle areas.

Land-augmenting mechanisation of this sort, combined with a shift to chemical fertilisers (to the gradual exclusion of manure), pesticides and weedicides, is undoubtedly labour-saving with respect to specific crops, but when the entire crop mix and enterprise combination is taken into account it may be labour-absorbing. However, as will be seen shortly, there are some States in India, where even the net result is labour-saving. In this context, it appears that efforts need to be made to measure the gains arising directly from land augmentation, as distinct from yield improvement with respect to specified crops, or crop combinations, in the case of labour-saving technological developments in particular. As yet, we often do not know why the labour-saving technology is being adopted in particular cases and specific regions.

In the long-run, in a land scarce economy, technological change which is land-augmenting is appropriate technology, even if the labour-saving properties are unwelcome. The evidence that such technological change is taking place in Indian agriculture suggests that we need to take a second, and more serious, look at policies for generating non-agricultural employment opportunities for workers now engaged in agriculture. While this could include policies to provide land and non-land productive assets to the rural poor, such efforts should be viewed as stop-gap measures. What most land-poor rural workers want is paid non-farm jobs, not petty self-employment. And this aspiration is in line with the present direction of changes in the rural employment structure, trends which are likely to get accentuated as time goes on. Increasingly, the share of the total workforce who are employers, self-employed direct producers, and tenants is tending to fall and correspondingly, the relative importance of hired workers to rise.

Thus the more practical policy may be to accelerate non-agricultural production growth rates, and to create conditions favourable to a higher elasticity of employment with respect to industrial and tertiary sector employment. In India, it appears that, aside from direct public investment and appropriate import policies, holding the product wage constant[17] in private (and public) industry, may be one of the most effective ways of doing this. In India, this would tend, first, to give a direct stimulus to employment, judging by Goldar's work; and secondly to enhance savings, investment and output growth in both the private and the public industrial sectors (judging by the work of Chakravarty[18] and Rangarajan,[19] among others) and hence to stimulate employment indirectly.

II. INDIAN EVIDENCE AT THE STATE LEVEL

Levels of Per Hectare Labour Absorption

In India, total labour absorption depends a great deal on the crop combinations grown, but there are certain States which can be identified as 'high' or 'low' labour absorption States for specific crops taken separately. For example, in all States, paddy, sugarcane, cotton and jute have relatively high labour requirements, but in some States two or three times as much labour is absorbed per hectare in paddy and sugarcane cultivation as in others.[20] In general, in the early 1980s, Haryana, Punjab, Madhya Pradesh and Rajasthan belong to the set of 'low labour absorption' States (65 man-days per hectare or less for all crops combined), while Andhra Pradesh, Bihar, Orissa, Tamil Nadu and West Bengal are characterised by high rates of per hectare labour absorption (more than 100 man-days per hectare). Among the intermediate labour intensity States, Gujarat is the most interesting case since on these criteria it belonged to the low labour absorption set in the early 1970s, with 62.22 man-days per hectare. Labour absorption in Gujarat has now risen to 86.16 man-days. This reflects the highest trend growth rate in per hectare labour absorption in India.

In all States cereals account for the largest share in total employment, ranging from a low of 57 per cent in Karnataka to a high of 91 per cent in Bihar. And in all cases where paddy dominants decisively in the production profile, it is found that labour absorption per hectare for 'all crops' is high. Low overall labour absorption per hectare, on the other hand, is peculiarly associated with crop cycles in which irrigated wheat is sown in the *rabi* season.

These inter-State contrasts are undoubtedly the product, in part, of differing soil-climatic conditions, and in part the outcome of contrasts in yield levels for specific crops. But this is not at all the whole story. As is abundantly clear from graphs, constrasts in labour intensity at any one point in time are also very much the cumulated product of the technological changes that have taken place over long, or even not so long, periods. The new technologies to varying degrees have been either labour-using or labour-saving, sometimes altering the relative positions of two States within the span of a single decade. These time profiles are the substance of the next sub-section.

Trends in Labour Absorption: An Overview

In December 1986, Vaidyanathan published an important paper[21] based on NSS data, indicating (among other things) the possibility that "in six states, though output rose, labour use in agriculture in 1977-78 was less than in 1972-73."[22] His figures are reproduced in col. (8) of Table 9.1. The six States were: Punjab, Uttar Pradesh, Haryana, Rajasthan, Madhya Pradesh, and Tamil Nadu.

These are the same States [marked with a star* in col. (6)] where the number of days of work available per male worker must have gone down, on the evidence of Cost of Production data, with the notable addition of Bihar, for which State Vaidyanathan finds an increase in employment. Moreover, on the basis of Cost of Production data, in

TABLE 9.1 : Trend Rates of Growth in Labour Absorption by States, 1971-72 to 1983-84 for 'All Crops' and Related Variables

	States (ranked according to col. (2) growth rates)	*Trend rates of growth in*				*Growth rates in*		*Per cent change in person-days employment 1977-78 over 1972-73 (NSS)*
		Labour absorption		*Gross cropped area under 'all crops'*	*Production (49 crops)*	*Male agricultural workers 1971-1981 (Census)*	*Labour productivity*	
		Total	*Per hectare*					
	1	2	3	4	5	6	7	8
1.	**Andhra Pradesh**	3.663***	2.779***	0.571	3.31	1.533	1.78	12
2.	**Gujarat**	3.574***	3.191***	0.371	3.92	1.544	2.38	14
3.	**Maharashtra**	3.435***	1.730***	1.676*	5.60	1.151	4.44	30
4.	**Karnataka**	2.614**	1.483***	1.114	2.44	1.691	0.75	9
5.	**West Bengal**	2.037*	2.096***	0.057	0.91	1.497	0.59	10
6.	**Rajasthan**	1.393	0.737	0.659	2.47	1.503*	0.97	–9
7.	**Orissa**	1.331*	0.795	0.531	2.28	1.126	1.15	4
8.	**Punjab**	1.079*	–0.887*	2.002***	3.92	1.287*	2.63	–2
9.	**Uttar Pradesh**	0.598	–0.145	0.743***	3.09	1.381*	1.72	–6
10.	**Haryana**	0.230	–0.357	0.589	3.31	1.841*	1.47	–11
11.	**Bihar**	0.074	1.037	–0.954	0.49	1.168*	–0.68	8
12.	**Madhya Pradesh**	–0.014	–0.976*	0.972***	1.65	1.619*	0.03	–1
13.	**Tamil Nadu**	–0.510	0.960*	–1.457*	1.12	0.858*	0.26	–13

Sources and Notes :

1. Cols. (2) and (3) : Computed from data gathered under the Comprehensive Scheme for Studying the Cost of Cultivation/Production of Principal Crops, for 18 crops, or fewer where data available for a smaller number of crops only.
2. Col. (4) : Data on gross cropped area is from Estimates of Area and Production of Principal Crops, Directorate of Economics and Statistics, Ministry of Agriculture, Government of India. Gross cropped area under 'all crops' relates to the 18 or fewer crops covered by the Cost of Production data.
3. Stars (*) in cols. (2) to (4) indicate t-value significant at the following levels : * 95 per cent, ** 98 per cent, *** 99 per cent.

4. Col. (5) : Trend growth rates in production for 49 crops were computed using indices of agricultural output for each State. The indices were prepared by the Commission for Agricultural Costs and Prices, and the data presented in G.S. Bhalla, *Some Issues in Agricultural Development in India: An Overview*, Table 3, in P.R. Brahmananda and V.R. Panchamukhi (Eds.) : *The Development Process and the Indian Economy*, Himalaya Publishing House, Bombay, 1987.
5. A star (*) in col. (6) marks those States where days of work available per male worker must have gone down, on the evidence of Cost of Production data on total labour absorption.
6. Col. (7) : Derived from data presented in Bhalla, *op. cit.*, Table 5, p. 248.
7. Col. (8) : *Source:* Vaidyanathan, *op. cit.*, *Economic and Political Weekly*, December 27, 1986, Table 12, p. A–141.

four of Vaidyanathan's six States, the trend rates of growth in per hectare labour absorption are negative for the longer period 1971-72 to 1983-84. In short, the identification of States with very low or negative rates of growth in labour absorption is roughly the same, on the basis of these two very different data sources.

As the figures in cols. (2) to (4) of Table 9.1 show, changes in gross cropped area (GCA) and cropping pattern shifts tend to offset to some degree the effects of low or negative trend rates of growth in per hectare labour absorption. The exceptions are Bihar, Tamil Nadu and West Bengal. In Bihar and Tamil Nadu, GCA under the crops covered appears to have gone down. In Tamil Nadu, it may be noted, the decline in total employment took place entirely because of the decline in GCA. West Bengal is a special case of another kind. Although both GCA and labour intensity have risen, it appears that cropping pattern shifts may have had a small adverse effect.

The net result of the combined effects of changes in labour intensity and changes in area and cropping patterns thus has been a rise in total employment in all States except Madhya Pradesh and Tamil Nadu. However, in seven of the 13 States, the growth of total labour absorption has been below the male workforce growth rate, and in six States, per hectare growth in employment is negative or not significantly different from zero.

However, by far the most interesting thing to emerge from Table 9.1 is the evidence of two very different paths of output-enhancing technical change.

Six of the 13 States are conspicuous for having production growth rates in excess of three per cent. And it is worth noting also that these are the only States which have enjoyed substantial improvements in per worker productivity. But these six States fall into two very distinct sets. On the one hand, Andhra Pradesh, Gujarat and Maharashtra have been adopting labour-using technologies to increase output (and yields). On the other hand, Punjab, Haryana and Uttar Pradesh have clearly shifted over to labour-saving technologies. In these States it has been possible to sustain existing levels of total employment only because of area expansion—(note especially the case of Punjab)—together with shifts of area to relatively more labour intensive crops. All told, however, technological change in these three States constitutes a package of basically healthy developments, as it has made it possible to improve per worker productivity substantially.

The remaining seven States all have much lower rates of growth in labour productivity. In Bihar and West Bengal the rate of growth of output per worker is in fact negative. But these States also fall into two sets. One set comprising Karnataka, West Bengal, Rajasthan and Orissa all have positive and generally significant growth rates in total man-days employment. The other three States, Bihar, Madhya Pradesh and Tamil Nadu, all suffer from negative or negligible growth rates in both total employment and per worker productivity. More disaggregated cropwise data are needed, however, to distinguish between those States (of the seven), where significant labour-using or labour-saving technological change is taking place in the cultivation of one or more crops, and those where it appears that nothing of note is happening at all, except perhaps the slow immiseration of people wholly or mainly dependent on agricultural labour for their livelihood.

Cropwise Trend Rates of Growth in Labour Absorption

The figures in Table 9.2 reveal that in some States man-days per hectare declined for all major crops, and further, that for some particular crops, mandays per hectare fell in nearly every State, regardless of what happened in the case of other crops.

Haryana, Madhya Pradesh, Punjab and Uttar Pradesh are the States where labour-saving technology now seems to dominate the entire spectrum of crops.

Among crops, wheat, barley and sugarcane generally show negative rates of growth in labour intensity. The only exceptions in the case of wheat are Bihar and West Bengal, and in the case of sugarcane, Maharashtra where there is a small, but significant positive trend rate of growth (0.95).

For the other major cereal crop, paddy, pet hectare labour absorption is rising in most of the traditional rice growing areas, but is falling in the main non-traditional rice producting States, Punjab and Haryana and also in Madhya Pradesh and Uttar Pradesh. These four are the States where growth in labour intensity is negative for almost all other crops. In addition, labour intensity in paddy cultivation has fallen substantially in Karnataka.

High rates of growth in labour absorption are associated with groundnut cultivation in the major groundnut producing States, and spectacular growth rates with recent shifts from traditional to superior

TABLE 9.2 : Cropwise Trend Rates of Growth in Labour Absorption Per Hectare, 1970-71 to 1984-85 by States

Crop	*Andhra Pradesh*	*Bihar*	*Haryana*	*Gujarat*	*Karnataka*	*Madhya Pradesh*	*Maharashtra*
1	2	3	4	5	6	7	8
Paddy	3.41***	1.12	–1.99***	(1.26***)	–2.89	–1.35*	(1.26***)
Jowar	3.27***	(2.08***)	–0.50***	3.03**	1.38	0.18	1.30
Bajra	2.58***	0.18***	–0.86	4.75***	1.67***	0.18***	1.84***
Maize	(0.71***)	0.61	–1.58***	1.09***	0.70***	–0.14***	(0.12***)
Wheat	–0.74***	1.52	–1.88	–1.31***	–0.74***	–1.17	–0.74***
Barley	—	0.71***	–1.14***	–1.53***	—	2.10***	–0.15***
Tur (*Arhar*)	0.14***	0.16***	–1.80***	0.14***	2.51	2.62	2.25
Moong	2.44	1.03***	–2.00***	—	0.63***	0.36	1.03***
Urad	0.82	0.19***	–2.13	—	0.20	0.56	0.91***
Gram	0.24***	(0.23***)	–2.64***	5.05***	0.24***	–1.26	(0.23***)
Groundnut	1.28	0.49***	(2.24***)	2.31	–0.33	2.17***	(2.24***)
Soyabean	—	—	—	0	—	0.83	—
Sunflower	0	0	—	—	0	—	0.03
Rapeseed and Mustard	(2.41***)	(2.41***)	–2.35	0	0	–0.26***	0
Cotton	–0.58	(1.51***)	–0.12	8.48***	9.21***	–3.63	2.02
Jute	—	–0.62	—	—	—	—	0
Sugarcane	–0.79	–3.07*	–1.23***	(–0.98***)	–0.92	(–0.98***)	0.95
V.F.C. Tobacco	–0.49	—	—	—	—	—	—
Paddy	2.14**	–2.36	(1.26***)	1.56**	–0.82	2.63**	
Jowar	0.83**	0.25**	2.49***	1.15***	–0.47**	(2.08***)	

TABLE 9.2 (*Contd.*)

Crop	*Orissa*	*Punjab*	*Rajasthan*	*Tamil Nadu*	*Uttar Pradesh*	*West Bengal*
1	9	10	11	12	13	14
Bajra	2.54***	4.34	2.14	4.27***	0.36**	2.56***
Maize	(0.12***)	4.61***	0.64	(0.71***)	0.05**	(0.70***)
Wheat	–0.74***	2.22***	–3.32***	–0.74***	–1.18	0.45
Barley	—	1.14**)	–2.26***	—	–1.42***	0
Tur (*Arhar*)	0	1.81***	0	0.01	0	0
Moong	–0.17	(–2.00***)	0.10	0	0	0
Urad	—	(–2.13***)	0	0.30***	0	0
Gram	–0.21***	2.53***	1.28	0	0.02	0
Groundnut	0.49***	(2.24***)	(2.24***)	–0.18	(2.24)	—
Soyabean	—	—	—	—	–0.00	—
Sunflower	0	—	—	0	0	0
Rapeseed and Mustard	(2.41***)	–1.69***	1.11	(2.41***)	1.07***	2.41***
Cotton	(1.51***)	0.19	2.13***	1.19**	(1.51***)	(0.51***)
Jute	0.70	—	—	—	0.53***	0.58
Sugarcane	(–0.59***)	(–0.98***)	(–0.98***)	–0.36	–1.08	(–0.59**)
V.F.C. Tobacco	—	—	—	—	—	—

Notes :

1. The 'man-hours per hectare' data upon which this table was based were rarely available for either of the two-end point years and were in no case available for all 15 years without a break. In most of the relatively complete time-series causes the data begin from 1971-72 and end in 1983-84, but have gaps for some intervening years. All data available for any of the 15 years (1970-71 to 1984-85) were used to calculate trend growth rates, making appropriate adjustments in the programme for any 'blank' years appearing between the triennia 1971-72 to 1973-74 and 1981-82 to 1983-84. Growth rates were not calculated at all unless reasonable estimates for man-days were available or could be generated for the six years comprising these two triennia.
2. Figures to four decimal places are available on request.

3. A '0' entry indicates that constant labour absorption per hectare has been assumed, in the absence of any time-series data or any firm basis for thinking that the figure for the early seventies should have been different from those of the late seventies or early eighties.
4. Stars: ***, ** and * indicate t-values significant at the 99, 98 and 95 per cent level respectively.
5. Brackets appear in the cases of some States for which no data whatsoever were available for certain crops, and yet rough estimates could be made for both triennia and basis of regional or national averages for States of similar yield levels for the specified crop and similar labour absorption behaviour in the case of other crops. The simplest example is that of the Haryana labour absorption data for barley, which was 'adopted' for Punjab in the absence of data for the latter State. These growth rate figures are presented here because the underlying estimates were an essential input in generating the total labour absorption figures presented elsewhere.

Details of assumptions underlying (bracketed) cropwise trend rates of growth in labour absorption per hectare given in Table 24.2.

1. *Paddy*: Based on assumed per hectare labour absorption figures computed from actual unweighted average labour absorption per hectare for roughly equal yield States. (Six States' data were used for the triennium 1971-74, and three for the trennium 1981-84).
2. *Jowar*: Based on assumed per hectare figures computed from actual national average (unwighted) data for each year of the triennia of the early 1970s and early 1980s.
3. *Maize:* Based on unweighted averages for States of similar labour absorption in other crops.
4. *Wheat:* Based on unweighted yearwise national average labour absorption figures for the two triennia.
5. *Barley:* Based on Haryana data.
6. *Moong:* Based on adopted Haryana figures.
7. *Urad:* Based on assumed fixed relation to labour absoption in other pulses in Punjab-Haryana.
8. *Gram:* Based on national averages.
9. *Groundnut*: Based on average of 'low labour absorption' States.
10. *Rapeseed and Mustard:* Based on average of medium to high labour absorption States with backward projection of Assam time-series data.
11. *Cotton:* Based on national average (unweighted) labour absorption figures.
12. *Sugarcane*: The (–0.9754) figure is based on the unweighted average of 'low labour absorption' States, and the (–0.5910) one on the unweighted average of 'high labour absorption' States.

varieties of cotton in Gujarat and Karnataka. However, Haryana and Punjab, which shifted to 'American' cotton some time ago, now exhibit negative or negligible changes in labour absorption.

In general, wheat seems to be a key element in the crop cycles of States characterised by a general decline in per hectare labour absorption. This suggests the following proposition for testing: in States where labour-saving technology was introduced during the past fifteen years or so, it was because it became advantageous when HYV wheat could be grown consecutively with one other remunerative crop in a two-season crops sequence. A supplementary proposition is: when labour-saving technology became advantageous with respect to one or two major crop sequences in a State (for example: wheat-rice, wheat-cotton, or sugarcane-wheat), then its adoption tended to push down labour absorption in other crops as well. The consequences in terms of the total employment available to agricultural workers need not be adverse, however, even with respect to the particular crops where labour intensity has declined.

In wheat, for example, in Haryana, Gujarat, Rajasthan and Uttar Pradesh, total employment has gone up, thanks to increases in area under this crop. The same is true of paddy in Haryana, Punjab and Uttar Pradesh, all of which recorded very rapid growth in total employment concurrently with declines in employment per hectare under paddy. Sugarcane in Gujarat and Karnataka also showed substantial gains in employment despite labour-saving technological change.

In short, as long as the spread of labour-saving technology is associated with major changes in cropping patterns in favour of *relatively* labour intensive crops, together with increases in gross cropped area, the total employment effect even with respect to particular crops may be positive.

But it would be foolhardy to predict that this state of affairs will continue. The favourable cropping pattern changes which have taken place and the increases in gross cropped area have been heavily contingent on public and private investment in irrigation. If the pace of irrigation expansion slackens, the story could be very different. It is possible that total agricultural employment in at least six States would go down during the coming decade. The States are: Rajasthan, Punjab, Uttar Pradesh, Haryana, Madhya Pradesh and Tamil Nadu.

Cropwise Elasticities of Employment with respect to Yield

For purposes of forecasting, it is convenient to have at hand estimates of the elasticity of per hectare employment with respect to yield. Such estimates are given in Table 9.3. However, a warning needs to be given at the outset. As in the case of trend rates of growth, so also in the case of these elasticities: given the inverted U-shape of the labour intensity trend lines in the case of several important crops, the elasticity estimates are not likely to yield very reliable predictions. For example, for States which have recently 'turned the corner' and are now moving down along the falling segment of the labour intensity trend line for a particular crop, the computed elasticities with respect to yield for such crops will be down right misleading. In fact, they will be positive when they 'should' be negative.

Thus it appears that while the computed State level cropwise employment elasticities presented here are revealing of past responses of labour intensity to changes in yield, by themselves they may not be much good for predicting future trends in labour absorption.[23] Judging by the contrasts between the 1971-72 to 1983-84 elasticity figures presented in Table 9.3, and the Tyagi estimates for the period spanning the triennium ended 1970-71 to the triennium ended 1978-79, there has been a major tendency, in crops other than jowar, for States to shift from the high positive to the low positive elasticity ranks and from the low positive to the negative elasticity sets.

In the case of several crops, graphically, the shift in the time span covered is tantamount to slicing an inverted U-shaped labour intensity graph further to the right, where AB is the typical 'Tyagi slice', and XY represents the time span covered by the present study.

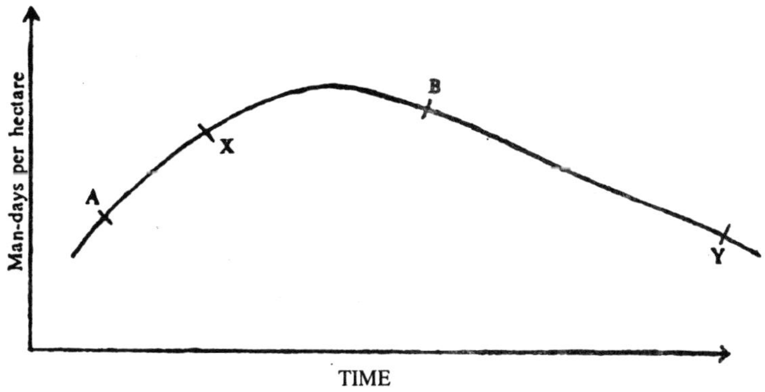

TABLE 9.3 : Per Hectare Employment Elasticities with respect to Yield for Specified Crops and States, 1971-72 to 1983-84

(Trend growth rate in man-days per hectare ÷ Trend growth rate in yield)

*Crop/(Tyagi's elasticity estimate in brackets)**	*Ranked negative elasticities*			*Low positive elasticities*			*High positive elasticities*			*'Labour sink' elasticites*		
	State	*Elasticity*	*Yield growth rate*	*State*	*Elasticity*	*Yield growth rate*	*State*	*Elasticity*	*Yield growth rate*	*State*	*Elasticity*	*Yield growth rate*
1	2			3			4			5		
1. Paddy (0.7265)	Haryana	–0.393	(5.04)	Karnataka	0.251	(1.14)	Tamil Nadu	0.904	(1.73)	West Bengal	5.599	(0.47)
	Punjab	–0.390	(6.04)	Gujarat	0.273	(4.63)	Andhra Pradesh	1.310	(2.60)	Orissa	23.722	(0.09)
	U.P.	–0.368	(2.23)	Maharasthra	0.329	(3.84)				Bihar	28.040	(0.04)
2. Wheat (0.7923)	Rajasthan	–1.194	(2.78)	None			Bihar	0.809	(1.89)	West Bengal	1.372	(0.33)
	Haryana	–0.953	(1.97)									
	Punjab	–0.890	(2.50)									
	M.P.	–0.592	(1.97)									
	U.P.	–0.488	(2.42)									
	Gujarat	–0.431	(3.04)									
3. Bajra (0.1501)	*Note:*—Negative or negligible growth rates for yield and production in States which are shifting out of bajra (Haryana, Punjab, Rajasthan, Madhya Pradesh, Uttar Pradesh), eliminated these States from the set for which elasticities were calculated.						Maharashtra	0.550	(3.35)	None		
							Andhra Pradesh	0.715	(3.61)			
							Tamil Nadu	1.482	(2.88)			
4. Jowar (0.4438)	Uttar Pradesh	–0.596	(0.79)	Maharashtra	0.244	(5.33)	M.P.	0.595	(0.31)	None		
				Karnataka	0.431	(3.21)	Tamil Nadu	0.741	(1.55)			
				Gujarat	0.435	(6.97)	Andhra Pradesh	1.397	(2.34)			

TABLE 9.3 (*Contd.*)

1	2	3	4	5
5. Maize	Madhya Pradesh –0.179 (0.77)	Bihar 0.436 (1.40)	None	None
6. Barley	Rajasthan –3.585 (0.63) U.P. –1.177 (1.21) Haryana –0.579 (1.97)	None	None	Bihar 1.538 (0.46)
7. Groundnut	*Note:*—A noteworthy feature of the underlying data is that in other States even when yield goes down, labour intensity has been rising.		Gujarat 0.622 (3.71) Andhra Pradesh 1.773 (0.72)	
8. Jute	U.P. –0.633 (0.84) Bihar –0.582 (1.07)	West Bengal 0.457 (1.28)	Orissa 0.894 (0.78)	None
9. Cotton	Haryana –0.107 (1.16) Andhra Pradesh –0.052 (11.22)	None	Rajasthan 0.538 (3.65) Maharashtra 0.918 (2.20) Tamil Nadu 1.033 (1.85) Karnataka 1.883 (4.89) Gujarat 6.471 (1.31)	*Note:*—Very high elasticities are associated with recent shifts from *desi* to Amercian cotton.
10. Sugarcane (0.3087)	U.P. –1.895 (0.57) Tamil Nadu –0.174 (2.04)	Maharashtra 0.315 (3.03)	*Note* :—Negative yield growth rates, combined with falling labour intensity, were recorded for all of the other States for which time-series cata were available, except Punjab, which had small negative yield growth rate and a small increase in labour intensity.	

*Tyagi's estimates are for the period from the triennium ending 1970-71 to the trennium ending 1978-79.

Some States with respectable yield growth rates, possibly for sound economic reasons, may be late in adopting technological packages which have enhanced incomes elsewhere, and hence continue to exhibit the relatively high elasticities associated with the ascending segment of the curve; other States may already be operating somewhat beyond the turning point, although the computed elasticities are positive.

Consider the case of paddy. Very similar negative elasticities appear for three of the six most dynamic States; Haryana, Punjab and Uttar Pradesh. For two other States with high yield growth rates (Gujarat and Maharashtra), the data generate low positive elasticities. At the other extreme, what might best be described as 'involution elasticities' appear for States with negligible long-term yield growth rates. In Bihar and West-Bengal, in particular, labour inputs have increased despite long-term stagnation in yields, declining productivity of labour in agriculture generally, combined with fewer days work per agricultural worker. This combination is appropriately described as 'labour sink', instead of as 'labour absorption', and the elasticities associated with this scenario are quite meaningless, since the increase in labour 'absorbed' was not in response to the requirements of a yield-raising technological change.

Clearly the factors that have caused these wide contrasts are forces which are liable to alter existing elasticities in future, possibly at short notice, States which have raised yields up to now through extending irrigation and drainage systems, using HYV seeds, more fertilisers and improved cultivation practices (to adopt Ishikawa's[24] list of labour-using technological factors), may move on to labour-saving technologies which enhance farm incomes by facilitating double cropping, or otherwise effectively augment the land available for crop cultivation. To predict this kind of switch, it appears that we need to know a great deal more about the reasons for past shifts from labour-using to labour-saving technologies.

All-India Elasticities of Employment with respect to Production: Specified Crops and Crop Categories

To derive an overview of the impact of state level changes in yield, in cropping patterns and in GCA combined upon total employment in crop producting, a regression was run in which the dependent variable was the set of state level trend rates of growth in total absorption, and the independent variable the statewise rates of growth in production. In the results presented in Table 9.4, the regression coefficients constitute a set of (unweighted) elasticities of employment with respect to the

Table 9.4 : Regresion Result: Dependent Variable: Statewise Trend Rates of Growth in Total Labour Absorption of Specified Crops: Independent Variable: Statewise Trend Rates of Growth in Production

	Crop/Crop group	*Regression coefficeint (Elasticity)*	R^2
1.	Paddy	0.9916 (–.3080)	0.3263
2.	Jowar	0.2278 (–.2454)	0.2535
3.	Bajra	1.0670 (–5886)	0.7819
4.	Maize	0.8872 (–1143)	0.7429
5.	Wheat	0.4473 (–.1453)	0.4353
6.	Barley	0.8143 (–0.741)	0.6530
7.	*Tur (Arhar)*	0.7554 (.0663)	0.6794
8.	Gram	1.4054 (.8318)	0.8028
9.	Groundnut	0.9928 (.2174)	0.7882
10.	Rapeseed and Mustard	0.9062 (-.1903)	0.6915
11.	Cotton	0.7997 (.0975)	0.3777
12.	Jute	1.4648 (.3233)	0.7273
13.	Sugarcane	0.8254 (.1542)	0.8535
14.	Cereals (except barley)	0.2597	0.4331
15.	Pulses and barley	0.9802	0.9682
16.	Oilseeds	0.8720	0.8431
17.	All crops (0.769)	0.5945	0.6034

Notes :

1. The figures in brackets in col. (2) are the corresponding all-India elasticities of employment per hectare with respect to yield.
2. The figure in brackets in item 17 is Tyagi's estimate for 'all crops'.

production of specified crops in India as a whole.

It is found that for all crops except jowar, the computed elasticities of employment with respect to production are much above the corresponding all-India elasticities of employment per hectare with respect to yield, given in brackets. More important, there are no negative elasticities of total employment with respect to production.

The corresponding R^2 values are also far higher in all cases, as the effects of dramatic inter-State contrasts in the impact of technological change on per hectare labour absorption in the production of certain crops are swamped by the generally positive employment impact of increases in area under such crops. R^2 are relatively low however for the key cereal crops: paddy, jowar and wheat, and for cotton.

Last but not least, it should be noted that one consequence of the sharp decline in the average per hectare employment elasticities for the key components of the cereal crop set is that the average elasticity with respect to production for 'all crops' estimated for the 1971-72 to 1983-84 period is considerably below the corresponding figure estimated by Tyagi (1981) for the ten- year period ending in 1978-79.

III. EQUITY IN THE DISTRIBUTION OF LAND AND NON-LAND PRODUCTIVE ASSETS

The Question of Equity in the context of Standard Change

In the economy as a whole, increasingly the share of the workforce who are self-employed direct producers, tenants, and/or employers is tending to fall, and correspondingly, the relative importance of hired workers to rise. At the same time, the share of workers of all sorts accounted for by agriculture in general is falling, and recent NSS data suggest that this rate of decline has accelerated.[25] In this context, the question of asset distribution among those who continue to own the means of production within agriculture takes on a set of somewhat different connotations that it has had up to now in the context of constant or worsening proportions in the workforce structure and rising absolute numbers mainly dependent on the land for their livelihood.

In the new context, the way we look at equity in asset distribution, in agriculture in particular and in rural areas generally, may need to be reassessed. Land is no longer the sole arbiter of rural destinies as it was in a 'state' agriculture-based economy. Nor is the ownership of non-farm business assets of the kind that can make a worker, or an entire household, self-employed. Today we need to take into explicit account alternative, paid, employment opportunities. Paid jobs, outside as well as inside agriculture, may effectively raise households towards the middle of the economic hierarchy and may do so more constructively than the acquisition of a pair of buffaloes, or a spot of land.

This leads to the question: why do people object to inequality in the distribution of productive assets anyway? For the sake of clarity, it is useful to distinguish, first, that inequality in asset distribution which stands in the way of increasing the productive capacity of scarce land and abundant labour (Table 9.5). Where that is the situation most economists would find in it a persuasive case for asset redistribution. Well and good. But suppose that the observed concentration of assets is, on the contrary, an integral part of a process which is increasing employment and labour productivity and making possible improvements in the standard of living of workers who have no such assets at all? Then what?

In that case people may still object because they find that in a particular rural setting, high or increasing inequality in the ownership of land and non-land assets is associated with increasing poverty among those endowed with few or no productive physical assets, or with the exercise of coercive power by the 'haves' over the 'have-nots' who are not in a position to claim a share in rising labour producitivy. In this case the question arises; is greater equity in the distribution of assets the best way to improve the position of the poor? Should the process of concentration of assets in the rural areas be discouraged, if and when it in fact takes place? I do not think there is any single, simple answer to these questions. I do think that we need to go beyond the Gini coefficients (or other measures of inequality) to identify what the processes are which have altered asset distribution (Table 9.5). In a dynamic, capitalist economy, while the association of increasing concentration in socio-political and economic power with asset inequality may be very great indeed, and persistent, it needs to be remembered that increasing asset concentration need not be associated either with increasing poverty or even with worsening income inequality. On the contrary, capital accumulation and its concentration are characteristic features of technologies which generate economies of scale and which increase labour productivity and the real incomes

TABLE 9.5 : Inequality in the Distribution of Land and Non-land Productive Assets by States, 1971, 1981 and 1972, 1982 (Gini Coefficients)

State/ Description	*Ownership holdings (base in hectares)*		*Specified assets (base in vlaue terms)*			
			Land		*Buldings*	
	1972	*1982*	*1971*	*1981*	*1971*	*1981*
1	2	3	4	5	6	7
I. High output growth: High labour absorption growth						
1. Andhra Pradhesh	.7668	.7016	.6481	.6037	.5292	.5133
2. Gujarat	.6441	.6343	.5792	.5955	.4481	.4670
3. Maharashtra	.6214	.6280	.5912	.5741	.4232	.4277
II. High output growth: Low labour absorption growth						
4. Punjab	.7589	.7498	.5391	.5990	.2808	.3500
5. Uttar Pradesh	.6144	.6194	.5931	.5696	.4324	.4732
6. Haryana	.7195	.6780	.5454	.3492	.3292	.2524
III. Low to medium output growth: Medium labour absorption growth						
7. Karnatka	.5946	.7497	.6297	.5698	.5815	.4934
8. West Bengal	.6371	.6372	.6005	.5806	.5117	.5181
9. Rajasthan	.5958	.5892	.6235	.5512	.4555	.4305
10. Orissa	.5997	.5828	.5882	.5943	.4582	.5191
IV. Low output growth: Negative or low labour absorption growth						
11. Bihar	.6737	.6731	.6664	.4745	.6264	.4932
12. Madhya Pradesh	.5826	.5013	.5894	.4442	.5916	.4408
13. Tamil Nadu	.7001	.7039	.6196	.5347	.6577	.5973

TABLE 9.5 (*Contd.*)

State/ Description	*Specified assets (base in value terms)*						
	Livestock		*Implements and machinery*	*Agricultural machinery*	*Transport equipment*	*'All assets'*	
	1971	*1981*	*1971*	*1981*	*1981*	*1971*	*1981*
1	8	9	10	11	12	13	14
I. High output growth: High labour absorption growth							
1. Andhra Pradhesh	.5126	.5025	.7210	.6881	.7536	.6192	.5765
2. Gujarat	.3855	.4430	.6110	.7033	.7430	.5341	.5483
3. Maharashtra	.4678	.4599	.6530	.6820	.7866	.5590	.5430
II. High output growth: Low labour absorption growth							
4. Punjab	.2139	.3400	.5214	.6549	.5623	.4933	.5589
5. Uttar Pradesh	.3868	.3694	.6321	.6770	.2956	.5487	.5397
6. Haryana	.2536	.2192	.5665	.5564	.4299	.4989	.3275
III Low to medium output growth: Medium labour absorption growth							
7. Karnatka	.4562	.4224	.6853	.6409	.8127	.6051	.4128
8. West Bengal	.4494	.4454	.6506	.6323	.7411	.5673	.5516
9. Rajasthan	.3666	.3413	.5542	.7017	.7035	.5475	.5069
10. Orissa	.4410	.4627	.5136	.5465	.7284	.5497	.5663
IV. Low output growth: Negative or low labour absorption growth							
11. Bihar	.4867	.4081	.7460	.7270	.8406	.6284	.5848
12. Madhya Pradesh	.4438	.4371	.5467	.6885	.6385	.5465	.5486
13. Tamil Nadu	.5021	.5111	.7631	.7338	.8482	.6021	.6272

Notes :

1. Basic data on rural ownership holdings are arranged by size class of ownership holdings, starting from 0.002-0.20 hectare.

(*Contd.*)

Sources :

1. "Survey on Land Holdings, National Sample Survey: Twenty-sixth Round (July, 1971-September, 1972), Parts I and II", *Sarvekshana,* Vol. V, Nos. 3 and 4, January-April 1982, and Government of India: National Sample Survey: Number 330—Thirty-seventh Round, January-December, 1982-Report on Land Holdings, National Sample Survey Organisation, Cabinet Secretariat, Ministry of Planning, New Delhi, March 1987.
2. Basic data on assets of rural cultivations are in value terms and are arranged by Household Asset Holdings Class defined in Rupees. 'Rural cultivators' include those having an operational holding on 0.002 hectare and above, as at 30 June, 1971 and 1981, respectively.

   Source: For 1971 for average value of specified assets and 'percentage of households reporting specified assets to all households in the households assets holding group', Reserve Bank of India: All-India Debt and Investment Survey, 1971-72-Assets of Rural Households, Bombay, 1976, Table 3.

   For 1981: derived from data in "A Note on Assets and Liabilities in Rural and Urban Households (States and All-India): National Sample Survey: Thirty-seventh Round (January-December 1982), Parts I and II", *Sarvekshana,* Vol. X, No. 1, July 1986, Tables 2 and 7.
3. Gini coefficients for 'All assets' relate to the productive assets listed above plus 'vacant house-sites' (in 1971 only) and 'non-farm business equipment' (1981 only) and exclude other assets covered by the NSSO-RBI Surveys.

of the working class.

In Indian agriculture, to the extent that mechanisation is a key element in the land-augmenting, yield-increasing, and labour-saving technological change which is taking place in some regions, one should in fact anticipate a period of worsening of asset distribution among the 'haves' and an increase in the number who 'own' only their labour power. The next sub-section examines only the first proposition.

Changes in the Distribution of Land and Non-Land Assets: The State Level Evidence

In brief, it appears that no remarkable changes in asset distribution have taken place over the past decade. But we cannot be sure because the basic data on specified assets currently available are arranged by household asset holding class defined in rupees, where the asset holding class relates to all assets, and not to specified productive assets. The correspondence with an underlying acreage class categorisation will be close, however, because of the dominance of the value of owned land in the determination of the asset holding class of most households.

The data available suggest that asset inequalities are at least no great in the three 'high technology' States than in others where not much labour saving mechanisation has taken place, despite the evidence that economies of scale are emerging in the advanced regions. More specifically, inequalities in Punjab, Haryana and Uttar Pradesh are no greater than in the high growth labour-saving technology States, and there is no well defined tendency for the coefficients to rise in these three 'high-tech' States. Gini coefficients in Punjab, Haryana and Uttar Pradesh are noticeably below those for all other States in respect of buildings, livestock and transport equipment.[26] The degree of inequality in the ownership of farm machinery is much higher than for other assets, however, and is in fact almost as great as in most other States. Moreover, in Punjab and Uttar Pradesh, it does appear that there may have been some increase in the degree of concentration during the last decade.''

Although we know that the really expensive labour-saving machinery is owned mainly by the big farmers, and that there are clear economies of scale in its ownership (although it is used by smaller farmers who purchase machine services extensively in the high-tech regions within 'labour-saving technology' States), it appears that many of the smaller farmers may have moved up towards the middle ranges of the machinery asset value categories through the recent purchase of an array of equipment other than the really big items. At least, this is one possible explanation for the remarkable stability of the Gini coefficients for machinery in the States where the ownership of major items of machinery is heavily concentrated among the large farm owners.[27]

On the basis of the NSS data currently available, however, not much more can be said. Clearly, a great deal of work needs to be done to find out in concrete terms exactly what has happened to non-land asset distribution in particular, and why. In the longer run we might expect a period of increasing concentration of owned assets as technologies providing economies of scale spread, provided that a larger subset of small and marginal farmers continue to cultivate their own land. However, to the extent that house-holds inheriting small plots of land turn to non-farm occupations for their livelihood, and sell their land, a regime of increasing equality could be established.

Note : Reproduced from *Indian Journal of Agricultural Economics*, Vol. 42, No. 4, Oct-Dec. 1987, pp. 537-560.

NOTES

1. See on Punjab, G.S. Bhalla and R.K. Sharma, ''Structural Changes in the Punjab Economy 1969-70 to 1979-80—A study in Input-Output Analysis (A Report on the International Food Policy Research Institute's Project)'', Centre for the Study of Regional Development, School of Social Sciences, Jawaharlal Nehru University, New Delhi 1987 (mimeo.) and on Uttar Pradesh, Sheikh Bhalla, ''Exercises in the Analysis of the Structure of Technological Change in Uttar Pradesh Agriculture'', Paper prepared for the seminar on Development Change in Uttar Pradesh, Sheila Bhalla, ''Exercises in the Analysis of the Structure of Technological Change in Uttar Pradesh Agriculture'' Paper prepared for the Seminar on Development and Change in Uttar Pradesh, March 27-29, 1987, Giri Institute of Development Studies, Lucknow (mimeo).
   In the case of wheat production, high R^2 values emerge for growth in the share of payments for fixed capitation value added in both States and in Uttar Pradesh for the share of Modern Intermediate Costs' (defined as the costs of Machine Labour and Fertilizers + Insecticides) in 'Total Intermediate Costs' (defined as the sum of the costs of bullock labour, machine labour, seed, fertiliser, manure, insecticies and irrigation charges). Lower R^2 and much lower trend growth rates emerge when irrigation charges are included as the fourth component in a second specification of 'Modern Intermediate Costs'.
2. Reference to 'all crops' in the present study relate only to those of the 18

crops covered by the Comprehensive Scheme for Studying the Costs of Principal Crops, for which State level data were available. The production trend growth rates used were calculated from indices of agricultural output prepared by the Commission for Agricultural Costs and Prices.

3. Tyagi's estimate covered eight crops, but for three of them the underlying employment (elasticity with respect to yield has been taken to be zero. See D.S. Tyagi, "Growth of Agricultural Output and Labour Absorption in India", *Journal of Development Studies,* Vol. 18, No. 1, October 1981, p. 112.
4. It is the weight of cereals in the total which pulls the 'all crops' elasticity down. Employment elasticities for specified crop groups are given below, together with R^2 values in brakcets. All crop 0.5945, (0.6034); cereals (except barley: 0.2597, (0.43317); pulses and barley; 0.9802, (0.9682) oil-seeds: 0.8720, (0.8431); sugarcane : 0.8254 (0- 8535); cotton: 0.7997, (0.3777); jute: 1-4648 (0.7273). The high jute figure appears to reflect area expansion and locational shifts in favour of relatively labour using areas. The State level lasticities with respect to yield are all less than one.
5. K.S. Krishnaswamy, "Issues for the Eighth Plan—I: What is New", *The Economic Times,* June 18, 1987.
6. Nicholas Kaldor: *Strategic Factors in Economic Development,* Cornell University Press, Ithaca, New York, 1967.
7. Simon Kuznets: *Economic Growth and Structure,* Oxford & IBH Publishing Co. Pvt. Ltd., New Delhi, 1965.
8. B. Goldar, "Employment Growth in Indian Industry", Paper presented as the Silver Jubilee Seminar of the Institute of Economic Growth, April 1984 (mimeo.)
9. Goldar uses the terms 'constant real wages' or constant 'real labour costs' interchangeably. Since the wholesale price index for manufactures is used as the deflator, in the aggregate across industries what he is getting is a product wage, which is constant in terms of its purchasing power over the products of industry at wholesale prices.
10. T.S. Papola (1987, as yet untitled) using data for the Annual Survey of Industries (ASI) sector Census plus Sample), covering the year 1968, and the years from 1975 to 1984 continuously, derives declining employment elasticities with respect to output as follows:

Period	*Elasticity of employment with respect to output*
1968-75	0.547
1968-84	0.422
1975-82	0.487
1975-84	0.287
1980-84	–0.006

While changes in 'real labour costs' (in aggregate, the product wage) were not taken into account separately in these calculations, it was noted that they had declined over the period.

11. Shigeru Ishikawa: *Essays on Technology, Employment and Institutions in*

Economics Development: Comparative Asian Experience, Kinokuniya, Tokyo, 1981, see especially pp. 187 and 239-240.

12. See Ishikawa: *op. cit.*, p. 37.
13. *ibid,* p. 36.
14. See Rohini Nayyar, ''Wages, Employment and Standard of Living of Agricultural Labourers in Uttar Pradesh'', in *Poverty and Landlessness in Rural Asia,* International Labour Office, Geneva 1977.
15. See Yujiro Hayami, ''Agrarian Problems of India from an East and South-East Asian Perspective'', Paper Prepared for a Seminar at the Institute of Economic Growth, Delhi, March 25, 1981 (mimeo.).
16. Hayami, *op. cit.,* p. 18.
17. How this could be done is beyond the scope of this paper, but it may be noted in relation to agriculture that the prices of wage goods, including foodgrains, should not be allowed to rise faster than the prices of industrially produced goods.
18. See S. Chakravarty, ''Reflections on the Growth Process in the Indian Economy'', in C.D. Wadhwa (Ed.): *Some Problems of India's Economic Policy: Selected Readings on Planning, Agriculture and Foreign Trade,* Tata McGraw-Hill, New Delhi, 1977.
19. See C. Rangarajan, ''Industrial Growth: Another Look'', *Economic and Political Weekly,* Vol. XVII, Nos. 14, 15 and 16, Annual Number, April 1982.
20. For the triennium 1981-82 to 1983-84, some of the extremes are listed below (in man-days per hectare).
    Paddy: Tamil Nadu 153.98, Andhra Pradesh 151.86, Madhya Pradesh 71.54, Punjab 91.09.
    Wheat: Bihar 92.03, Madhya Pradesh 45.62, Punjab 49.44, Haryana 51.15.
    Groundnut: Tamil Nadu 105.88, Gujrat 68.55.
    Cotton: Andhra Pradesh 173.28, Punjab 94.42.
    Sugarcane: Maharasthra 383.44, Andhra Pradesh 343.89, Tamil Nadu 341.91, Haryana 132.43, Uttar Pradesh 139.20.
21. A. Vaidyanathan, ''Labour Use in Rural India: A Study of Spatial and Temporal Variations'', *Economic and Political Weekly,* Vol. XXI, No. 52, December 27, 1986.
22. *Ibid,* p. A-137.
23. Ideally, one should identify in advance the turning points (if any) for each crop in each State and then calculate two separate elasticities one for the rising trend line and another for the declining portion. Unfortunately, some more years will have to elapse before sufficient data become available for most crops to make this kind of exercise feasible.
24. See Ishikawa: *op. cit.,* p. 36.
25. See *Sarvekshana,* Vol. IX, No. 4, April 1986, Table 3, p. S-112. It may be noted, however, that within agriculture, to the Cost of Production data, the

share of total labour days accounted for by family labour is rising. Labour-saving technology may tend to accelerate the pace of this development.

26. In Punjab and Haryana, the same was true for house-sites in 1971.
27. See for example, G.K. Chadha: *The State and Rural Economic Transformation: The Case of Punjab, 1950-85,* Sage Publications India Pvt. Ltd., New Delhi, 1986, especially Chapter 7.

10

LABOUR USE IN RURAL INDIA

A Study of Spatial and Temporal Variations

A. VAIDYANATHAN

I. SALIENT FEATURES OF LABOUR USE IN RURAL INDIA

For a long time the decennial population censuses were the only source of information on the pattern of employment in rural India: Every census gives some idea of the proportion of rural population engaged in economic activity, and their distribution by sector of employment. Pretty detailed sectoral break-up of the working population is generally available upto the state level though at the district level only broad sectoral pattern of employment is published. However, problems of determining the activity status of particular individuals, the fact that workers often do several types of work covering different 'sectors', and the difficulties of capturing the intensity of employment in a short questionnaire administered at a single point of time have severely limited the usefulness of their data. In order to overcome these problems, the NSS took up intensive surveys of a scientifically selected sample of the population from rural and urban areas all over India. Though these surveys were started in the fifties, it took time to evolve suitable standardised concepts and procedures—a process which was greatly facilitated by researchers who sought to understand the various facets of manpower utilisation by using the NSS and other data. (See especially the Dantwala Committee Report 1972 and papers presented at a seminar held in Centre for Development Studies, 1975). This process was largely completed by the early 1970s. Since that time that NSS has not only been careful to maintain comparability of design, concepts and procedures, but has also published rather detailed tabulations which give a fuller and more meaningful picture of the composition of the labour force, level and pattern of employment and incidence of unemployment.

Rural Labour in 1977-78 : A National Profile

In 1977-78 the estimated working population—persons usually engaged in productive activity[1]—in rural India was around 198 million; including the usually (i.e. chronically) unemployed, the labour force (in terms of usual status) was around 201 million. The labour force consisted of 139 million males and 61 million females representing about 65 per cent and 39 per cent respectively of the total population of the two sexes aged 5 years or more.

More than 60 per cent of both male and female workers in rural areas were self-employed or employed in family enterprises; the remainder were wage labourers. Somewhat more than one-fourth of the male wage labourers were working on regular wage salaried employment, the rest being casual labourers. Most (over 90 per cent) of female wage labourers were in the latter category.

In terms of their usual status about four-fifths of the rural male workers[2] and 86 per cent of females were employed in agriculture (including animal husbandry, fisheries and forestry). Mining and manufacturing (including utilities) absorbed a little over of 7 per cent of male workers (6 per cent among female); and trade, transport and finance (5.5 per cent among 2 per cent females); and services 5.1 per cent (3 per cent among females).

Not all people classified as 'usual workers' are employed all the time. A better measure of employment would be in terms of the number of days worked; on an average day, only 56.1 per cent of males actually had work while 64 per cent were categorised as workers according to their usual status. The difference is even more marked in the case of females where the comparable proportions are 22 per cent and 38 per cent respectively. Also the extent of unemployment measured with respect to the daily status is considerably higher than with reference to usual status; while barely one per cent of the male workers (according to usual status) report themselves to be chronically unemployed, unemployment measured on the basis of current daily status works out to over 7 per cent of the labour force.

Altogether in 1977-78, an estimated 62.7 billion persondays were spent on economic activity;[3] reported unemployment was 4.7 billion man days. Roughly four-fifths of the total workdays are spent in agriculture—78 per cent among males and 83 per cent among females. A little over 70 per cent of the total employment (in terms of person days) is accounted for by males two thirds represents the contribution of self-employed (own account workers and their family helpers). Of the remainder, which measures the extent of wage paid employment, almost three-fourths (72 per cent) are accounted by casual wage labour. The incidence of wage employment relative to total employment is higher in non-agricultural sectors (48 per cent compared to 30 per cent in agriculture); over half of rural non-agricultural wage employment is on a regular basis while over 80 per cent of wage employment in agriculture consists of casual labour. The incidence of casual wage employment is also much greater among females (Table 10.1). The data also suggest that while most of the casual non-agricultural employment is in manual work, about half of regular wage employment is non-manual work.[4]

TABLE 10.1 : Pattern of Labour Use in Rural India 1977-78

(10^9 days/annum)

	All Activities				*Agriculture*			
	Family	*Regular Employees*	*Casual Labour*	*Total*	*Family*	*Regular Employees*	*Casual Labour*	*Total*
Male	30.2	5.3	9.7	45.3	25.2	2.4	7.8	35.4
Female	11.0	0.8	5.6	17.3	9.3	0.3	4.8	14.4
Total	41.2	6.1	15.3	62.6	34.5	2.7	12.6	49.8

Source : Sarvekshana, J-O 1981: S37-S38.

The NSS also gives some idea of the break-down of agricultural work by major operations. These data, summarised in Table 10.2, shows ploughing, sowing, transplanting, weeding and harvesting—which are all important operations in crop production— account for barely two-fifths of the total labour use (in terms of persondays) in this sector. About 5 per cent is used for non-manual work—presumably supervisions. But over 55 per cent of the labour input is classified as 'other manual work': A finer breakdown of this category would be useful to get a proper appreciation of the importance of other operations related to crop production, and of activities other than crop production (such as animal husbandry).[5]

TABLE 10.2 : Distribution of Employment (in Persondays) in Agriculture by Operations, Categories of Labour and Sex, All India 1977-78

Operations	*Per cent Distribution of Total Labour Input*			*Ratio of Female to Male Labour by Type of Labour*			
	Male	*Female*	*Total*	*All*	*Family Labour*	*Casual Labour*	*Regular Workers*
Ploughing	14.0	1.5	10.3	0.043	0.04	0.05	0.003
Sowing	2.0	1.6	1.9	0.332	0.27	0.61	0.11
Transplantation	2.7	5.8	3.6	0.863	0.50	1.52	0.45
Weeding	7.2	14.7	9.7	0.831	0.66	1.32	0.22
Harvesting	12.7	19.5	14.7	0.629	0.52	0.90	0.23
Other manual work in agriculture	55.9	53.8	55.8	0.393	0.40	0.47	0.11
Non-manual work in agriculture	5.5	3.1	4.8	0.227	0.23	0.32	0.08
Total	100	100	100	0.408	0.37	0.62	0.11

Source : Sarvekshana, J-O 1981: S37-S38.

There are also marked differences in the relative importance of female labour between different operations: They figure prominently in transplantation, weeding and harvesting: in fact they out-number the males in the casual labour category in the latter two operations, but are relatively less important or play a minor role in others (notably ploughing and non-manual work).

Changes Over Time

The NSS also provides the estimated population and its employment characteristics on the basis of sample data for 1972-73, 1977-78 and 1983. They suggest that the labour force participation rates in terms of usual status of males aged 5 or more years has remained more or less stable during this period, while that of females shows a slight increase in the latter part of the period compared to 1972-73. However, the workforce participation rates in terms of current daily status—which are invariably lower—show a progressive decline in the case of males; a fall between 1972-73 and 1977-1978 and stability thereafter in the case of female[6] (Table 10.3). This would seem to suggest average duration of employment per usually employed worker (measured by the ratio of CDS to US participation rates) has fallen over this period; but almost all of it occurred between 1972-73 and 1977-78. This index does not show any change—if anything there is a marginal improvement—in 1983 compared to 1977-78.

TABLE 10.3 : Labour Force Participation Rates* in Rural India

	1972-73		*1977-78*		*1983*	
	US	*CDS*	*US*	*CDS*	*US*	*CDS*
Male	64.6	58.0	64.9	56.1	64.2	55.4
Female	37.7	25.2	39.3	21.6	39.1	21.8

Notes: * Per cent of population in each sex aged 5 or more years who are in the labour force (= working population + unemployed). Includes main and marginal workers. US = Usual Status; CDS = Current Daily Status

Source : GOI, NSSO (1985).

There has been a progressive and substantial decline in the proportion of male workers in the self-employment category. Among females, the proportion fell in 1977-78—the fall being somewhat less pronounced than among males—but did not change thereafter. The obverse of this phenomenon is the progressive rise in the share or wage labour in the workforce—from 34 per cent to 40 per cent in the case of males and 36 per cent to 38 per cent among females. Another significant trend is the growing casualisation of wage labour, the tendency being again more pronounced among males: male casual wage labourers as a proportion of male wage labourers rose from 64 per cent in 1972-73 to 73 per cent in 1983. Female wage labour, predominantly (88 per cent) casual to being with, has also increased reaching 92 per cent in 1983 (Table 10.4).

TABLE 10.4 : Composition of Workers (Main and Marginal) According to Usual Status by Category of Employment

(*Percentages*)

	1972-73		*1977-78*		*1983*	
	M	*F*	*M*	*F*	*M*	*F*
Self-employment	65.9	64.5	62.8	62.1	60.4	62.2
Regular wage/salaried employment	12.1	4.1	10.6	2.8	10.8	3.1
Casual wage labour	22.0	31.4	26.7	35.1	28.8	34.7
Total	100	100	100	100	100	100

Source : GOI, NSS (1985): 12.

It is also possible to get a rough idea of the trends in employment in terms of person days of work: Between 1972-73 and 1977-78, total employment of persons between the ages of 15 and 59 years is estimated to have risen by only 2.3 per cent (from 51.6 billion person-days to 52.8 billion person days)—male employment rising by some 7 per cent and female employment falling by about 9 per cent. On the other hand, based on estimates for population aged 5 or more years, total employment as well as that of males and females taken separately seems to have remained at the same level as in 1977-78 (Table 10.5). However, the change between 1977-78 and 1983 should be interpreted with caution because (a) the 1983 estimates are based on data for only two sub-rounds (covering six months) while 1977-78 estimates relate to the full year; (b) the rise in rural population between 1977-78 and 1983 on the basis of NSS estimates seems to be substantially below the rate shown by the census.

TABLE 10.5: Estimated Total Employment by Sex and Sector, India 1972-73 to 1983

(10^9 person days)

	1972-73[1]			*1977-78*			*1983*[4]		
	M	*F*	*T*	*M*	*F*	*T*	*M*	*F*	*T*
Agriculture	27.6	12.7	40.3	29.7	12.3	42.0[2]			
				35.4	14.4	49.8[3]	33.2	13.6	46.8
Non-agriculture	8.2	3.1	11.3	8.5	2.3	10.8[2]			
				9.9	2.9	12.8[3]	12.1	3.6	15.7
Total	35.8	15.8	51.6	38.2	14.6	52.8[2]			
				45.3	17.3	62.6[3]	45.4	17.2	62.5

Sources :

1. *Sarvekshana,* January 1980: (relates to workers in age group 15-59. The concept of workers corresponds to main marginal workers).
2. *Sarvekshana,* April 1979: (relates to main workers in age group 15-59).
3. *Sarvekshana,* July-October 1981: 537-538 relates to main workers aged 5 years or more.
4. GOI, NSS (1985) : relates to workers (main-marginal) aged 5+years. Based on estimates for January to June 1983.

The pattern of rural employment shows a significant and sustained tendency to diversification. The percentage of rural workers (in terms of their usual status) reporting to be employed in agriculture and related activities has fallen among both males and females, the fall among males (from 83.2 per cent in 1972-73 to 76.8 per cent in 1983) being more pronounced than among females. The increase in the share of non-agricultural sectors is distributed widely among different categories of non-agricultural activity, though the gain in construction, transport, trade and services seems to be relatively more than in manufacturing (Table 10.6). We do not have as detailed sectoral distribution of person days of employment but we can get some idea of the trend in the relative growth of agricultural and non-agricultural employment from Table 10.3. This gives a slightly different picture in that the share of non-agricultural sector in total employment fell between 1972-73 and 1977-78 (the fall being quite pronounced among females), but had risen substantially thereafter especially among males (22 per cent to 27 per cent). Indications are that the share of non-agricultural sector in total employment (person- days) in 1983 is significantly higher than in 1972-73.

TABLE 10.6 : Sectoral distribution of Main and Marginal Workers (According to Usual Status) 1972-73 to 1983

(*Percentages*)

	1972-73		*1977-78*		*1983*	
	M	*F*	*M*	*F*	*M*	*F*
Agricultural hunting, forestry and fishing	83.2	89.7	80.6	88.1	76.8	86.7
Mining and quarrying, manufactur ing electricity gas and water	6.3	4.9	7.1	6.2	8.0	7.2
Construction	1.6	1.0	1.7	0.6	2.6	0.9
Wholesale and retail trade, transport, storage and communication financing real estate, etc.	4.2	1.5	5.5	2.0	6.5	2.1
Community social and persona services and others	4.7	2.9	5.1	3.0	5.9	3.0
Total	100.0	100.0	100.0	100.0	100.0	100.0

Source : GOI, NSS (1985): 14.

Regional Variations

There are large regional variations in practically every facet of labour utilisation. Salient features of these variations based on 1977-78 data can be seen from Table 10.7. The overall labour force participation rate (in terms of usual status) for males varies from 55 per cent in Kerala to over 70 per cent in Andhra Pradesh; and for females from 13-14 per cent in Assam to 54 per cent in Rajasthan. In terms of current daily status, the participation rate are generally smaller, and the range of variation much larger. (These statistics are not presented in Table 10.7) (2) the extent of open involuntary unemployment (in terms of CDS) ranges from a low 2 per cent of labour force in MP to nearly a third in Kerala. (3) Wage employment as a proportion of total rural employment, which is less than 10 per cent in Rajasthan and Jammu and Kashmir, exceeds 40 per cent in Kerala, Tamil Nadu, Orissa and West Bengal. Wage employment is largely in the form of casual labour in most states, but in a few—notably Punjab, Haryana and Assam—the major part of it is accounted by regular employment. (4) The share of non-agricultural activities in total employment ranges from barely 12 per cent in MP, to nearly 30 per cent in West Bengal and an exceptional 49 per cent in Kerala. (5) The quantum of employment in agriculture relative to the area sown also shows large variations.

TABLE 10.7 : Estimated Employment* and its Components in Rural Areas by States 1977-78

State	*Total Employment*	*Employment in Agriculture*			*Wage Labour*		*Casual Labour as Per cent of Wage Labour*[3]
		Total[1]	*Per Ha*[2] *NSA*	*(3) as Percent of (2)*	*Total*[1]	*(6) as Per cent of (2)*	
1	2	3	4	5	6	7	8
Andhra Pradesh	5.96	4.58	420	76.8	2.50	35.6	64.6
Assam	1.60	1.35	499	84.5	0.69	33.2	49.0
Bihar	6.19	5.03	585	81.2	2.43	39.1	77.5
Gujarat	2.96	2.53	266	85.3	0.95	28.3	74.3
Haryana	1.03	0.80	215	77.7	0.24	15.4	44.2
Jammu and Kashmir	0.48	0.36	414	74.2	0.09	5.0	40.4
Karnataka	3.48	2.97	280	85.4	1.39	37.5	75.5
Kerala	1.84	0.94	428	51.2	0.87	42.2	57.1
Madhya Pradesh	6.02	5.25	281	87.4	1.81	27.2	68.5
Maharashtra	5.96	4.79	262	80.3	2.63	39.7	55.6
Orissa	2.70	2.14	356	79.3	1.11	40.8	68.4
Punjab	1.38	1.08	239	78.3	0.43	25.0	45.5
Rajasthan	3.72	3.18	209	75.4	0.48	7.5	57.7
Tamil Nadu	4.40	3.18	506	72.3	1.51	42.3	63.1
Uttar Pradesh	10.25	8.30	450	81.0	1.85	15.2	61.7
West Bengal	4.07	2.91	452	71.4	1.83	42.3	64.8

Notes :

* Relate to main workers among males and females aged 5 years or more.

1. 10^9 days per annum.
2. Person-days.
3. Males, current daily status.

Source : Computed from NSS estimates for the 32nd round, *Sarvekshana*, April-June 1981.

There are also significant differences between states in terms of the direction and rate of change in the pattern of labour use over time. Table 10.8, which summarises the available data for three points of time over the last decade, shows:

1. Between 1972 and 1977 employment grew at a slower pace than rural population in all but 2 states. This was also the case in several states during the subsequent five years, but in a majority of them, employment growth kept pace with or even exceeded population growth. Overall, in Andhra Pradesh, Bihar, Maharashtra, Rajasthan and Tamil Nadu employment growth seems to have persistently fallen short of population growth over the period, the gap being particularly pronounced in Tamil Nadu.
2. In a majority of cases unemployment rates fell between 1972 and 1977 but in the subsequent six years, it shows a rise in a majority of cases. Only Tamil Nadu shows a persistent rise in unemployment rate in both periods—the 1983 level being 60 per cent higher than in 1972. A few states (AP, Bihar, MP) have recorded a progressive fall.
3. The share of non-agricultural employment rose in a majority of states—the rise being particularly marked in Haryana, Punjab, Jammu and Kashmir, Karnataka, Tamil Nadu and West Bengal.
4. The incidence of wage labour shows a progressive rise in a majority of states (11 out of 16)—the rise in the first half of the period being in general more pronounced than in the latter part. Interestingly there are several cases where the incidence of wage labour has remained stable (Kerala) or even declined (Maharashtra).

In what follows we intend to explore further the factors underlying the inter-state variations in respect of the labour use in agriculture and the importance of non-agriculture employment. The focus will be primarily on variations at a point of time, but to the extent possible we will also examine the differences in respect of direction and magnitudes of changes over time.

II. LABOUR USE IN AGRICULTURE

The NSS estimates of agricultural employment imply that in 1977-78 on the average 355 days of human labour (of all ages and sex) were used per hectare of net cultivated area and close to 300 days per hectare of gross cropped area. At first sight this seems quite out of line with

TABLE 10.8 : Changes in Selected Aspects of Rural Male Employment in Major States, 1972-83

	Employment[1]			*Unemployment*[1]			*Non-Agricultural Employment*[2]			*Wage Labour*[3]		
	1972-73	*1977-78*	*1983*	*1972-73*	*1977-78*	*1983*	*1972-73*	*1977-78*	*1983*	*1972-73*	*1977-78*	*1983*
	(1)	(2)	(3)	(4)	(5)	(6)	(7)	(8)	(9)	(10)	(11)	(12)
Andhra Pradesh	63.5	61.8	60.4	6.9	5.7	5.6	23.5	25.3 24.1	23.9	40.0	45.8	46.6
Assam	55.7	53.0	53.6	1.5	0.8	1.9	18.5	9.4 15.1	21.9	23.8	32.3	35.9
Bihar	55.7	53.9	52.9	5.7	5.7	4.2	20.2	18.9 19.3	22.2	39.5	40.9	42.8
Gujarat	57.3	55.6	55.6	3.5	4.5	3.0	21.4	18.3 17.5	22.4	34.4	37.2	40.4
Haryana	56.5	52.4	55.1	2.2	4.0	3.4	20.6	26.0 23.5	26.8	24.0	28.2	30.0
Jammu and Kashmir	55.5	56.4	50.5	5.2	4.9	8.4	19.4	25.6 23.7	35.8	9.4	20.3	26.8
Karnataka	59.2	58.4	58.4	4.6	4.5	4.7	20.5	19.2 18.6	28.3	37.8	41.8	41.5
Kerala	41.1	38.4	40.6	12.8	13.7	13.4	50.3	49.5 47.8	47.9	54.3	53.8	54.5
Madhya Pradesh	61.5	59.7	60.7	2.2	1.9	1.6	12.5	13.4 13.8	16.6	27.3	31.0	33.7
Maharashtra	58.4	58.2	56.0	5.6	3.4	4.0	34.8	24.0 23.9	29.0	53.5	48.7	47.4
Orissa	58.3	55.8	57.5	6.3	4.5	5.1	22.4	20.6 19.6	28.2	39.9	45.6	43.7

TABLE 10.8 (*Contd.*)

	(1)	(2)	(3)	(4)	(5)	(6)	(7)	(8)	(9)	(10)	(11)	(12)
Punjab	60.8	58.6	59.4	2.9	2.9	4.1	19.9	25.2				
								23.0	26.8	30.7	36.0	33.6
Rajasthan	64.2	60.5	57.1	3.2	2.6	3.1	21.7	27.3				
								19.7	24.2	10.2	17.1	20.5
Tamil Nadu	61.6	57.5	52.7	7.2	8.4	12.0	24.9	31.5				
								30.1	39.0	42.4	48.4	54.2
Uttar Pradesh	58.7	55.7	56.4	1.9	2.5	2.2	20.4	21.5				
								20.7	24.4	21.6	24.4	24.1
West Bengal	52.4	54.5	50.7	7.2	5.7	8.8	25.9	26.7				
								26.5	33.2	48.1	49.0	50.4

Note : It should be noted that the 1972-73 figures roughly correspond to the category of main *plus* marginal workers introduced in 1977-78. The figures for 1977-78 in respect of 15-59 years relates to main workers only; that for 5+ years in 1977-78 and 1983 relate to main + marginal workers.

1. GOI, NSS (1985) : Key Results of the Last 3 Quinquennial NSS Enquiries on Employment and Unemployment Report No 315 (Delhi, mimeo) figures refer to percentage of population aged 5 years and more in terms of current daily status during January-June of each year.
2. The top row, against each state gives percentage non-agricultural to total employment for males in the age group 15-59 years. The bottom row gives the percentage for males aged 5+ years. All figures refer to CDS and are from *Sarvekshana* January 1980 (for 1972-73) *Sarvekshana* April 1979 from 1977-78 (15-59), *Sarvekshana* January-April 1981 for 1977-78 (5+ years) and GOI NSS (1985) Report No. 315 for 1983.
3. Computed from *Sarvekshana* October 1977 (for 1972-73); January-April 1981 (for 1977-78), and Report No. 305 of NSS for 1983).

estimates derived from other surveys, notably the Farm Management Surveys (FMS) conducted during the 1950s and the 1960s. The FMS data, covering some 20-25 districts, give a figure varying from 55 man days per hectare of gross crop area to a maximum of 190 days.

However, it needs to be noted that the FMS got information from the farmers regarding the quantum of labour used in specified operations connected with crop production and animal husbandry. The NSS on the other hand asked the sample workers as to the kind of work they did and the amount of time they spent on it. The FMS listing of operations is not exhaustive of all activities related to crop production—there being ambiguities concerning post-harvest operations and the notion of time spent on a particular operation. (For instance how is travel time between job to be treated?) Their coverage of animal husbandry is incomplete in that they provide estimates only for maintenance of work animals. There can also be differences between the perception of the employers and the workers in reckoning 'working time'. In general, the NSS would seem to be more comprehensive in its coverage of activities and may be expected to give somewhat higher figures of labour input than the FMS type of enquiry.

A second factor is that the estimates based on FMS cited above are in terms of standard mandays in order to allow for differences in the quantum (and quality) of labour effectively put in by workers of different ages and sex. It is conventional to postulate that females and children effectively do only a fraction of the work done by adult males. NSS estimates do not make any adjustment for age-sex differences in the quantum of work per unit of time. For both these reasons, it is not surprising that the NSS estimate do not make any adjustment for age-sex differences in the quantum of work per unit of time. For both these reasons, it is not surprising that the NSS estimate of labour input should be higher than the estimates from FMS. We will be using both sets of data in the subsequent discussion for both show large variations in labour intensity of agriculture across regions.

Table 10.9 presents state-wise estimate of total labour input into agriculture (and related activities) per unit of land in 1977-78 along with data on related characteristics having a bearing on labour use. Table 10.10 gives similar data for selected districts from different parts of the country on the basis of the FMS[7].

As noted earlier the average labour input per hectare of net sown area according to NSS estimates ranges from around 210 persondays

TABLE 10.9 : Labour Use in Agriculture and Related Characteristics by Major States, 1977-78

	HLD/NSA	*Average Rainfall*	*IA/NSA*	*IAW/IA*	*Tractors per 000 ha*	*Wage Labour / Total Labour in Agriculture*	*Cropping Intensity*	*Foodgrains Yield Per/Ha*	*Per cent Area Labour Intensive Crops*	*Nutrients Per/Ha nsa*
Andhra Pradesh	420	90	32	21	1.63	39	1.15	1.43	31.9	47.8
Assam	499	210	21	neg	0.76	28	1.24	0.97	19.7	2.2
Bihar	585	126	35	38	1.69	37	1.35	0.87	19.9	20.9
Gujarat	266	87	18	78	2.66	27	1.09	1.07	3.1	30.4
Haryana	215	61	52	41	13.97	16	1.49	1.39	10.4	51.8
Jammu and Kashmir	414	183	42	1	1.96	6	1.37	1.53	26.7	16.8
Karnataka	280	84	14	26	1.29	34	1.08	0.98	9.4	26.3
Kerala	428	281	10	18	1.10	37	1.32	1.54	11.3	35.4
Madhya Pradesh	281	114	12	40	1.15	26	1.14	0.72	4.3	8.5
Maharashtra	262	87	10	58	1.27	38	1.09	0.74	4.2	19.7
Orissa	356	145	19	9	0.43	38	1.32	0.91	16.8	10.7
Punjab	239	59	78	57	21.75	25	1.53	1.94	15.4	108.6
Rajasthan	209	50	19	56	2.19	8	1.12	0.58	0.1	7.4
Tamil Nadu	506	99	46	34	2.03	38	1.23	1.52	37.4	67.8
Uttar Pradesh	450	104	51	58	4.58	15	1.34	1.19	13.0	49.3
West Bengal	452	188	27	1	0.48	41	1.27	1.07	19.7	27.8

in Rajasthan/Haryana to nearly 600 days in Bihar. The FMS data which relate to districts dispersed all over the country show the labour input for crop production above to vary from less than 50 standardised man-days per hectare of net sown area in Ahmednagar to 200 days in Surat/Bulsar (Table 10.10). Including labour used for maintenance of draught animals raises the figure by around 10 per cent, but does not significantly affect the extent of variations in labour intensity of agriculture.

The existence of large variations in labour intensity of agriculture across Asian countries, and the possible reasons therefore were examined by Ishikawa in his pioneering work published in 1967. The problem was further explored, this time in a more detailed way and with greater emphasis on intra-country variations, in a series of studies under the auspices of the ILO-ARTEP. This series included three studies relating to India—one (Alagh *et al*, 1978) dealing with inter-district variations in labour intensity defined in terms of the stock of male labour relative to cultivated land; a second (Bardhan, 1978) focusing on inter-farm variations in labour input in selected districts; and the third (Vaidyanathan, 1978) seeking to explain variations in the number of labour days per unit area across selected districts and across crops and operations. Our focus here is on inter-state and inter-district variations in total labour input into agriculture per unit area and the reasons therefore.

Ishikawa (1967, 1978), it may be recalled, sought to unravel the role of technological and institutional factors affecting labour intensity: He distinguished between land augmenting technological change which raises the output per unit of land and thereby facilitates greater use of labour and changes (typically associated with mechanisation of farm operations) leading to a saving in the amount of human labour needed for cultivation. He argued that in the initial stages of development, technical change of the former variety is more prominent and results in a progressive rise in labour input per unit area. At a more advanced stage the influence of labour saving technical changes becomes more important, leading eventually to a decline in labour intensity of agriculture. He saw irrigation as the crucial element—or the 'leading input' as he called it—of land augmenting technical change in as much it is a pre-condition for other land augmenting changes (such as higher cropping intensity and more intensive use of fertiliser) to take root.

TABLE 10.10 : Basic Data Relating to Total Crop Production and Associated Variables in Selected Districts

District	Survey Period	Value of Output in Rs.	Per cent of Area Irrigated	Per cent of Irrigated Area under Wells	Per cent of Area under Paddy and Sugar-cane	Human Labour in Days (HLD)	Bullock Labour in Days (BLD)	Mechanical Power in HP	Cost Ratio of HLD/BLD	Hired Labour as Per cent of HLD	Cropping Intensity	Total Rainfall in mms	Rainfall in June-Sept. as Per cent of Total Rainfall
(1)	(2)	(3)	(4)	(5)	(6)	(7)	(8)	(9)	(10)	(11)	(12)	(13)	(14)
Ahmednagar	1954-56	265	18.5	28.6	1.0	44.3	19.2	0.014	0.81	32	104.8	574	77
Nasik	1954-56	299	12.8	26.0	0.2	56.1	25.4	0.017	0.77	29	101.8	1059	77
Pali	1962-64	303	24.0	54.2	0.1	65.7	22.3	0.016	0.39	na	102.4	369	96
Surat Bulsar	1966-68	817	18.3	23.0	23.5	168.5	11.2	0.041	0.10	60	118.0	2906	99
Amritsar/ Ferozepur	1954-56	838	74.0	54.0	1.1	51.0	22.1	0.080	0.51	27	129.0	693	86
Samabalpur	1957-59	894	10.4	24.0	85.5	106.7	30.0	0.004	0.60	42	97.0	1300	92
Cuddappah	1967-69	1075	35.4	48.0	18.8	85.3	19.5	0.088	0.49	68	93.0	697	49
24-Parganas	1954-56	1137	2.0	0.1	81.0	112.0	17.8	0.006	0.78	33	106.6	1490	75
Meerut/ Muzaffarnagar	1954-56	1153	72.9	24.7	23.5	93.2	29.6	0.031	0.23	32	134.4	1105	80
Cuttack	1967-69	1210	50.6	11.5	73.7	101.2	14.2	0.003	1.07	40	140.4	1401	82
Hooghly	1954-56	1299	25.3	47.4	64.0	146.6	20.0	0.021	1.02	42	106.0	1420	75
West Godavari (paddy zone)	1957-59	1371	75.5	2.8	70.5	103.0	15.7	0.003	0.24	79	156.0	1195	63
Deoria	1965-68	1551	30.1	97.8	29.8	100.9	24.9	0.064	0.28	41	128.5	1022	77

TABLE 10.10 (*Contd.*)

(1)	(2)	(3)	(4)	(5)	(6)	(7)	(8)	(9)	(10)	(11)	(12)	(13)	(14)
Ferozepur	1967-69	1622	87.0	51.7	4.9	60.0	13.5	0.125	0.46	49	129.0	327	90
Coimbatore	1970-72	1743	50.0	42.0	16.0	93.4	9.1	0.299	0.33	85	143.0	773	28
Thanjayur	1967-69	1818	91.0	1.3	82.6	118.8	24.1	0.030	0.97	75	151.0	1229	30
West Godavari (tobacco zone)	1957-59	1871	19.9	37.7	25.7	145.0	16.1	0.003	0.33	78	199.0	1092	66
Muzaffarnagar	1966-68	2006	92.0	38.6	32.3	73.2	19.1	0.142	0.21	40	139.0	860	87

Notes :

1. Unless otherwise indicated, all figures are three-year averages per hectare of gross cropped area.
2. Output is evaluated, uniformly at all-India average price for individual crops in the period 1970-72.
3. HLD = Human Labour Days; BLD = Bullock pair days; HP = Horsepower equivalent of engines; CI Cropping intensity.

Source : Vaidyanathan (1978).

Among the institutional factors Ishikawa stressed were the size of holding, the incidence of tenancy and the working of the market for inputs and outputs. That there is a marked inverse relation between labour intensity of cultivation and size of holding has been noted much earlier and different explanations offered (Sen, 1962; Srinivasan, 1971). Ishikawa argued that in the absence of outside opportunities, involuntary unemployment was likely to be more under 'market economy arrangement' than under community type arrangements (characteristic of Japan in earlier times), feudal landlords with monopsonistic or oligopsonistic power with respect to labour is recognised as a third type of institutional arrangement with distinctive consequences for labour intensity. Where farming is carried on by cultivator families, he argued, if land lease and/or labour market were sufficiently competitive and the supply of land for rent/demand for wage labour are reasonably elastic, there need not be much of an employment problem. If these conditions are not met, "disguised unemployment is likely to emerge under market economy arrangements though it can be avoided to the extent that community arrangement for providing employment to all members are in operation in some form" (Ishikawa, 1978: 89-91).

Bardhan's empirical analysis of the variation in labour intensity across farms in Bengal also emphasises differences in technology and institutions irrigation, multiple cropping, fertilisers and HYV are the major elements of technology considered in his analysis; among institutional factors average farm size, percentage of area leased in, bargaining power of labourers and the wage rate for hired labour are prominent (Bardhan, 1978).

Our data set which relates to 16 states and 18 districts do not permit a detailed analysis of the impact of institutional factors, partly for the reason that relevant information is not available and partly because the observations are far too few. We will therefore focus here primarily on the role of technological and environmental factors in explaining the inter-regional variations in labour intensity of agriculture, with some observations on the changes in labour intensity in recent years. Our analysis suggests that while technological factors are important, differences in agro-climatic environment are a significant factor influencing the productivity of land and hence the intensity of labour use.

Let us begin with a relatively simple question, namely, whether the level of labour input per unit area bears any systematic relation to the productivity per unit area. Both the state cross-section and the district cross-section show that there is a significant positive correlation between the two. That is states/districts where the output unit of gross cropped area is high tend to use a longer input of human labour per unit of gross area. Data for different size class of holding in a few selected districts—the choice being determined by availability of data—also show in most cases a positive association between the two variables but the correlation is statistically significant only in nine out of 13 cases (Table 10.11). In three districts labour input per unit area trends to fall as output per unit area rises, the tendency begin significant in all cases. The elasticity of labour input relative to productivity per unit of cropped area taking only cases where the coefficient of association is significantly different from zero—ranges from –0.5 to nearly +1.1, the majority lying in the range 0.75 to 1.0. But the small number of observations available in this data set and the fact that they relate to group averages makes them a rather unreliable basis for a confident inference. This will have to wait the analysis of farm level data. In the rest of this section we will be concerned with only the variations in the averages across states and the selected districts.

TABLE 10.11 :Elasticity of Labour Intensity in Crop Production With Reference to Output Per Unit Area in Selected Districts

District	*Elasticity*	*Number of Observation*	*Significance of Elasticity Coefficient*
Ahmednagar	.995	8	S
Surat-Bulsar	–.347	5	S
West Godavari (paddy zone)	.770	8	S
Meerut/Muzaffarnagar	.493	5	S
Nasik	.945	8	S
Sambalpur	.813	5	S
Raipur	1.093	5	S
Pali	.598	8	S
Muzaffarnagar	.196	5	NS
Thanjavur	–.243	5	S
Deoria	.245	4	NS
Cuttack	.825	5	S
Ferozepur	-.514	5	S
Cuddappah	.105	5	S
Amritsar/Ferozepur	.697	5	NS
Coimbatore	.813	5	S

Note : Estimated on the basis of mean values of output and labour input unit gross cropped area for different size class of farms as reported in the various Farm Management Surveys. The estimating equation is log (HLD/GCA) = a + log (output/GCA).

TABLE 10.12 :Estimated Change in Employment and Output in Agriculture, 1972-73 to 1977-78

State	*1972-73*		*1977-78*		*Per cent Change 1977-78 over 1972-73 in*	
	Employ-ment 10^6 PD[1]	*Index of Output*[2]	*Employ-ment 10^6 PD*[1]	*Index of Output*[2]	*Employ-ment*	*Output*
Andhra Pradesh	3374	143	3781	188	12	31
Assam	1079	144	1302	146	21	1
Bihar	4139	146	4474	160	8	10
Gujarat	1967	166	2242	352	14	112
Haryana	727	284	641	384	–11	35
Karnataka	2168	194	2363	290	9	49
Kerala	764	185	834	182	9	–1
Madhya Pradesh	4600	147	4560	176	–1	20
Maharashtra	3158	88	4106	228	30	159
Orissa	1791	153	1799	179	4	16
Punjab	880	284	865	384	–2	35
Rajasthan	2656	189	2416	276	–9	46
Tamil Nadu	3101	236	2693	265	–13	12
UP	7044	160	6618	196	–6	22
West Bengal	2244	136	2471	183	10	34

Notes :

1 Relates to 15-59 years, males and females.

2. A V Jose (1982).

It is useful to start with proximate factors influencing the overall labour intensity of cultivation: Variations in labour per cropped hectare must be seen as the combined effect of variation across space/farms in the quantum of labour used to raise each individual crop, and in the proportion of area allocated to different crops. The latter, which reflects differences in crop pattern, is important particularly in the Indian context because (a) wide variety of crops are raised and (b) their labour requirements vary widely. It is therefore, reasonable to expect that, other things being equal a region/farm where the crop yields are higher will tend to use more labour; and that where labour intensive crops are more important overall labour intensity is likely to be higher. One can think of ways of constructing indices to capture the pure yield differences and the pure crop pattern differences across the regions/farms as the case may be, but we have not attempted it in the present study. Instead we have used the overall yield per hectare of cropped area[8] and the proportion of area devoted to labour intensive crops as approximation of the two factors[9]. But for any given level of productivity and crop pattern, the extent of mechanisation of cultivation operations could make a significant difference to the labour intensity. If we were to explain variation in labour input per unit of cultivated land (as distinct from cropped area) it would be necessary to bring in besides the above three factors, differences in cropping intensity as well. The relation may be written thus

$$\frac{HLD}{GCA} = f\left(\frac{O}{GCA}, \% \, LIC, \frac{TP}{na}\right)$$

$$\frac{HLD}{NSA} = \frac{HLD}{GCA} \times \frac{GCA}{NSA}$$

where

HLD = human labour days
O = output
GCA = gross crop area
TP = Tractor power
$\%LIC$ = percentage of GCA devoted to labour intensive crops.

The above relations were estimated with a linear multiple regression for both the inter-state cross-section (Table A).

TABLE A : Coefficients of Multiple Regression of Labour Intensity of Agriculture

	O/GCA	*Per cent LIC*	*TP/HA*	$\bar{R}^2$
Inter-state cross-section	–17.6	5.611	–7.683	.67
	(–.273)	(3.068)	(–2.193)	
Inter district	.018	.446	–99.44	.22
	(1.201)	(1.641)	(–.775)	

The inter-state cross-section corroborates that the proportion of area devoted to labour intensive crops has a strong positive relation to overall labour intensity of agriculture, while tractorisation has a depressing effect on labour absorption. The output per gross cropped acre by itself does not seem to have significant bearing on labour input per unit of cropped area. In the inter-district cross-section, somewhat surprisingly, none of the three factors seem to have a significant association with labour intensity, though the signs of the coefficients are in the expected direction.

A different way of approaching the problem and one which has the merit of focusing on the factors underlying variations in cropping intensity, yield and crop pattern would be to relate the differences in labour intensity to differences in respect of agroclimatic environment and technology, keeping a distinction between those elements of technology which are labour-augmenting and those which tend to displace labour. Much of the available analysis of labour use in agriculture focuses on the effect of technology. There are numerous studies on the effect of particular types of technology (irrigation, HYV, mechanisation) on labour intensity (Mehra 1976; Bartsch, 1977).[10]

There is general agreement that the elements of bio-chemical technology (like irrigation, HYV, fertilisers) are all land augmenting and therefore conducive to absorption of more labour. But there is difference of opinion on whether mechanisation of operations other than irrigation are, on balance, labour absorbing or labour saving. Some people (including Ishikawa) hold that mechanisation by increasing speed of operations makes it possible to complete timely sowings on a larger area and to raise intensity of cropping above the levels feasible with manual or animal-drawn equipment. Both these being labour absorbing may more than compensate for the reduction of labour use in the operation which is mechanised. The weight of evidence in India (Agarwal, 1980; Binswanger, 1978), however suggests that mechanisation is distinctly labour saving and we will work on that hypothesis.

Irrigation, as Ishikawa so strongly emphasised, is in many ways the crucial element of bio-chemical technology: The scope for using other inputs (notably HYV and fertilisers) and their productivity depends critically on the availability of soil moisture and the timeliness and assurance of its supply. And irrigation is supposed to make an improvement over rainfed cultivation in some or all these respects, besides making a significant difference to crop pattern. But quality of irrigation varies and the proportion of area irrigated, which is conventionally used as the irrigation index, is not sufficiently refined to capture differences in quality of irrigation. Insofar as groundwater irrigation is superior to surface irrigation in terms of the above criteria, the ratio of area irrigated by wells to total irrigated area could be used as an index of irrigation quality. We will also do the same, but it must be noted that this is till not wholly satisfactory inasmuch as it does not capture differences in the quality of surface irrigation (in terms of quantum and assurance of water supply) or of differences in the extent of conjunctive use of surface and ground water. The extent of area receiving irrigation is also not a satisfactory index of soil moisture. Part of the reason is that the area measure fails to discriminate between wide differences in the quantum, timing and assurance of water supplies.

But more importantly, since soil receives moisture both from irrigation and rainfall, the effect of the former on cropping intensity, crop pattern and productivity depends significantly on the quantum and seasonal distribution of rainfall. (For a more detailed discussion *see* Vaidyanathan, 1985.) Ideally we need to have a measure of the soil moisture satuts in different seasons reflecting the combined effect of rainfall and irrigation. Since this is not possible—essentially because we have no data on the volume of water delivered from different irrigation sources by season or of the irrigation efficiency—we have to make do with the area measure of irrigation but in conjunction with information on rainfall.

Insofar as the level of use and productivity of other bio-chemical inputs (especially HYV and fertilisers) are a function of soil moisture, we need not take explicit account of the latter. The complementarity between soil moisture and the level of other bio-chemical inputs and their productivity, however is not strict: Variations in these respects can and do occur within any given moisture regime for a variety of other reasons, e.g., soil conditions, knowledge of farmers, relative prices. Therefore, there is some point—subject to data availability—to including fertiliser inputs proportion of area under HYV, per cent of area treated against pests and disease as independent factors affecting productivity and hence labour use.

The other major technological determinant of labour use given bio-chemical inputs, in the degree of mechanisation of farm operations (other than irrigation, which is clearly labour augmenting and incorporated in the irrigation index). The effect of mechanisation of particular operations on overall cropping intensity or productivity cannot be settled at the macro level. We can only see whether the degree of overall mechanisation (excluding irrigation) has, given the bio-chemical inputs, a net positive or negative association with labour intensity. We use the total horsepower of the tractor stock as the index of mechanisation.

$$\frac{HLD}{NSA} = \frac{HLD}{GCA} \cdot \frac{GCA}{NSA}$$

$$HLD/GCA = f\,(O/GCA,\ \%\ LIC,\ TP/NSA)$$

$$\text{since } \frac{GCA}{NSA} = f_1\,(NIA/NSA, \frac{IAW}{IA}, TR, RJS)$$

$$\frac{O}{GCA} = f_2 \frac{GIA}{(CCA, F, HYV, TP)}$$

$$\% \, LIC = F_3 f_3 \frac{NIA}{NSA}, \frac{IAW}{IA}, TR, RJS$$

we may view

$\frac{HLD}{NSA}$ as a function of

$\frac{NIA}{NSA}, \frac{IAW}{IA}, TR, RJS, \frac{TP}{NSA}, F$ and HYV.

IAW = Irrigated area by wells
TR = Total Rainfall
RJS = % rainfall in June-Sept.

In the present exercise we have not considered F (because it tends to be highly correlated with irrigation and crop pattern) and HYV (because reliable data are not available).

A linear multiple regression analysis of the inter-state cross-section shows that other things being given regions with higher total rainfall tend to have a higher labour intensity: the higher the proportion of irrigated area the higher the labour intensity; and the greater the extent of tractor use the less is the labour intensity. All these relations are statistically significant. There is but a weak positive association between the proportion or irrigated area served by wells and labour intensity. The analysis of inter-district cross-section confirms that there is a significant positive association between labour intensity on the one hand and total rainfall, and (somewhat less strongly) the extent of irrigation. It also shows that the greater the seasonal concentration of rainfall the lower the labour intensity. Mechanisation, however, does not have a significant relation with the labour intensity (Table B).

TABLE B : Coefficients of Multiple Regression of Labour Intensity of Agriculture (HLD/NSA) on Selected Variables

	TR	*Per cent RJS*	*Per cent IA*	$\frac{IAW}{IA}$	$\frac{TP}{NSA}$	$\bar{R}^2$
Inter-state	0.965 (2.609)		6.902 (4.726)	.454 (.465)	–23.14 (4.58)	.723
Inter district	.073 (5.603)	–1.879 (–2.871)	.511 (2.147)	.464 (1.359)	–134.98 (–1.199)	.67

Note : Figures in brackets are T values.

Our data confirm that the overall irrigation ratio does indeed have significant influence on labour intensity of agriculture, but it can explain only a small proportion of the spatial variations in the latter. Climatic variables and specifically rainfall and its seasonal distribution turn out to be important factors.

As mentioned earlier we do not have sufficient information or large enough body of observations to explore the role of institutional factors in determining labour intensity of agriculture. However the data permit us to see whether there is any systematic relation between labour intensity and the extent of dependence on wage labour—the latter being an indicator of the relative roles of cultivation using wholly or mostly family labour and cultivation depending largely or wholly on wage labour. Taking the cue from Sen's explanation for the inverse relation between farm size and labour intensity[11] we would expect that the higher the dependence on wage labour, the lower *ceteris paribus*, will be the labour intensity. But, as the estimates of multiple regression given below show, other things being the same, states/districts with a higher proportion of wage labour in total labour input into agriculture tend to have *higher* total labour input per unit of land, the relation is statistically significant in the case of the state cross-section but not in the data relating to selected districts. A more disaggregated analysis of data at least at the regional level—which should be possible if the regional level tabulations of NSS data were available—would be useful to test the robustness of this as well as other relations concerning labour intensity (Table C).

TABLE C : Multiple Regression of Labour Intensity of Agriculture on Selected Variables

	TR	*RJS*	*Per cent IA*	*TP/ha*	*HL/AL*	$\bar{R}^2$
Inter-state	0.772 (3.479)		7.115 (6.437)	−22.007 (5.183)	3.117 (2.676)	.829
Inter district	.062 (4.616)	−.358 (.732)	.483 (2.16)	−53.51 (.558)	.698 (1.738)	.693

Note : Figures in brackets are T values.

Variations Over Time

The picture regarding changes in labour intensity is rather confused: We saw earlier that there is in general a positive association between output per hectare and labour input per hectare across farms in a given region, and that the elasticity of labour input relative to output (per unit area) in general exceeds 0.5. But evidence of changes in labour intensity in a given area over time is rather scanty.

Comparable data on productivity of land and quantum of labour input per unit area at different point of time are readily available only for four districts from the Farm Management Surveys. They suggest that labour inputs per hectare has risen at a much slower rate than output per hectare and that in some cases have actually declined (Vaidyanathan 1978; Sinha 1979). Some limited analysis of data from cost of cultivation data also suggest similar trends. Cross-section data clearly do not give a reliable basis for judging the relation between yield improvement and labour use in a dynamic context.

Alagh and Bhalla have put together some data at the district level and found that the rate of growth of rural male workers tends to be higher in districts with relatively high rates of output growth and argued that this reflects and differential demand for labour associated with differential growth of output. While this association is significant and suggestive, several points need further probing before firm inferences can be drawn: First, the comparison is based on averages for only four categories of districts grouped according to growth rate of output. Second, those experiencing relative stagnation or decline in output do not fit the pattern in that the growth of male worker population in this category is much higher than in low growth district. Third, the data relate only to male workers, and that too to the stock of labour power without any allowance for differences in the intensity of its use. Clearly this offers a rich area for further work.

Data on the quantum of labour used in agriculture available from the NSS suggest that for rural India as a whole labour use in 1977-78 was only some 4 per cent higher than in 1972-73 while crop production was at least 20 per cent higher.[12] At the state level, while total agricultural employment generally rose, the rise was well below the growth of output in all cases, except Assam and Kerala; in six states though output rose, labour use in agriculture in 1977-78 was less than in 1972-73. there is no relation between the percentage change in agricultural employment and percentage change in output.

Now there are reasons to expect employment in agriculture to grow slower than output: we have already seen that the elasticity of employment relative to yield per hectare based on the experience of a few districts is substantially below unity and yield increases have been a significant source of growth in recent years. At a more general level, while certain types of labour using technical changes (notably the spread of irrigation and of ground water in particular) have taken place, this may be in some measure counter balanced by the rapid mechanisation of groundwater irrigation at least in so far as employment in irrigation is concerned. The fact that the ratio of farm yard menure (which is highly labour intensive) to chemical fertilisers (which involves handing smaller volumes of material per unit nutrient) has fallen steeply implies that increase in labour input in fertiliser and manure application would have been much slower than the growth in the quantum of nutrients applied. And mechanisation of other farm operations (including ploughing, harvesting, threshing and transport) are clearly labour saving.

What the balance of these influences has been in different parts of the country and whether the changes in labour use in agriculture as shown by NSS are a reasonably accurate reflection thereof are questions which we are not in a position to answer definitively. Part of the reason is that there is some basis to believe that the NSS may be under-estimating the real increase in agricultural, employment in as much as, in the context of growing specialisation of activity, the reporting of sectoral distribution of work and workers may be getting more and more discriminating. (See Section III for further discussion of this point).

III. RURAL NON-AGRICULTURAL EMPLOYMENT

While agriculture is unquestionably the dominant source of employment in rural India, non-agricultural activities are by no means inconsequential: At the time of the 1981 census, nearly a fifth of the rural workforce (main workers). were reported to be employed in non-agricultural pursuits. The National Sample Survey (NSS) estimates that in 1977-78, in terms of their usual satuts, nearly 20 per cent of male workers and around 12 per cent of female workers in rural India are employed in non-agricultural activities. In terms of mandays of actual work, non-agricultural activities accounted for 22 per cent of total employment in the case of males and one sixth in the case of females.[13] In absolute terms the volume of non-agricultural employment in rural areas accounts for nearly a third of the total non-agricultural employment in the country. A serious study of the determinants of rural non-agricultural employment is therefore in order.

There exists quite large regional variations in ratio of non-agricultural to total employment in rural areas: According to the 1981 census, this ratio (in respect of main workers) ranges from around 12.6 per cent in MP to over 25 per cent in West Bengal and an exceptional 52 per cent in Kerala. According to NSS (32nd round) data, the ratio of non-agricultural to total employment in rural areas (in terms of man-days) varied from 12 per cent in MP to 30 per cent in West Bengal and again an exceptional 50 per cent in Kerala. An exploration of the reasons unerlying these variations could hopefully give us some basis to judge how far they are amendable to manipulation by policy and through what policies.[14]

Factors Influencing Rural Non-Agricultural Employment

Broadly speaking the level of rural employment in non-agricultural activity (relative to the rural labour-force) can be viewed as a function of the following factors: (1) the level of rural demand (relative to population) for various non-agricultural goods and services produced locally; (2) the level of extra-local demand—from urban areas in the vicinity as well as from other regions—for rural products (and services) (3) location, scale and technology of activities catering to these demands.

Demand Factors—Local Rural Demand

The rural sector's demand for non-agricultural products consists of inputs for agriculture and animal husbandry, manufactured goods for final consumption and/or capital formation, consumer services of various kinds and trading, transport and related services.

Traditional agriculture uses mostly locally produced inputs, a large part of it derived from agriculture and animal husbandry. The few non-agricultural inputs—such as marking and repair of implements—were also produced by local artisans (usually under the *Jajmani* system). With introduction of modern technology—e.g., fertilisers pesticides, and mechanical equipment like pumps and tractors—the dependence on non-local inputs increases greatly. Most of the modern inputs, except perhaps machine repair services, are produced by industries located outside, often far away from the villages where they are used. Consequently, while the growth of agriculture production generally accounts in a more than proportionate increase in the demand, for inputs from non-agricultural input, it does not necessarily lead to an increase in demand for locally made inputs taken as a whole.

The final demand for manufactures and services in a given rural area is in part a function of the level of prosperity of its population, of whom agriculturists are the most important segment. Given that these are highly income elastic demands, it is to be expected that the rural demand for manufactures and services relative to population will be higher in areas where the per capita income of agriculturists is high and/or the distribution of incomes in more uneven. However, not all this demand is more uneven. However, not all this demand is for goods and services produced in the village. The relative size of the demand for local vs outside products is in part of a function of consumer preferences which in turn may be systematically related to per capita incomes—richer people wanting newer, more exotic products and products of superior quality. The extent to which the consumers are exposed to, and have access to, modern, urban products and life styles—which in turn depends on the extent of linkages with outside markets—could also affect the preferences as between local as against outside manufactures. (Linkages with markets outside the village also affects production and employment in rural areas—via the relative economic advantages of different locations and techniques—on which more later.)

The rural demand for personal and community services in a given area—which are mostly locally produced—again seems likely to be a function of the general prosperity of the rural population in that area. However there are certain categories of community services—e.g., public administration, education and modern health services—whose level and location may be determined exogenously, e.g., by state policy. These have expanded very rapidly in the last three decades and there is some evidence that rural employment under this category has become quite sizeable in recent years.[15] The level of state expenditure in rural areas obviously has second round effects on the demand for products of rural non-agricultural sectors and hence non-agricultural employment.

In the case of trade, transport and finance the level of local demand relative to population is related to per capita incomes (of which the volume of production per head is clearly an important determining factor) and the extent to which the local rural economy is commercialised. The larger the output per capita and the proportion of output which is sold outside, the larger will be the volume of trade and trade-related activity. That a high degree of commercialisation in production also implies a greater degree of dependence on market purchases for intermediate and final goods adds to the volume of commercial and transport activity.

Extra Local Demand

The market for products of rural non-agricultural activity is in many cases clearly confined to be local villages. This is particularly true of trade, transport, and services. But it is not always and necessarily so: Several Products of rural industry—e.g., earthen ware, leather footwear, processed farm produce like sugar and textiles—are sold in neighbouring urban markets and sometimes over a much wide area. Little is known about the extent to which rural industries products are meant for extra-local demand. However, if we are considering non-agricultural activity as a whole, and more so if we exclude manufacturing, it may not be unreasonable to assume that the 'markets' are mostly local.

It could also happen that some part of the rural workers reporting non-agricultural employment may be employed in nearby urban areas but residing in the village. To this extent a part of the non-agricultural workers residing in rural areas may in fact be working outside the catering to extra-local demands. How large a component this category is depends on the proximity to urban centres, the size of the urban centres relative to the rural population of the area, the case of access to transport and the costs of commuting. The larger the nearby towns, shorter the commuting distance and cheaper the cost the larger is likely to be the incidence of this phenomenon. We know little about where and how important this phenomenon is.

Organisation, Scale and Technology

For any given level of demand, the level of employment could vary depending on what kinds of techniques are being used, and how production is organised. The number of persons employed to process a given volume of say paddy will clearly by much smaller with modern rice mills than with traditional hand pounding. Whether non-agricultural activity is specialised or is an adjunct to agriculture may also make a difference to the employment, at any rate as reported in response to survey type enquiries.

The degree of commercialisation of the rural economy would seem to be a major factor affecting the scale, location and technique used in rural non-agricultural activity. Traditionally village communities were *relatively* self-sufficient and had an internal division of labour based largely on caste and regulated by the *jajmani* system. Trade and transport activity was limited, and most of the transport was done by bullock carts or on animals. The level of development of the transport network, as well as the size and density of village settlements would seem to be relevant in determining the location, organisation and hence labour intensity of non-agricultural activities.

The spread of commercialisation alters the picture in several important respects: First, agricultural production gets more specialised and the extent of trading—and trade related activity—increases. Second, increased exposure to and contact with nearby urban market centres may alter the pattern of rural demand for non-agricultural products in favour of extra-local resources. Third, since commercialisation of agriculture extends the territorial network of exchange, the effective size of the market for non-agricultural goods and services also expands, creating thereby the conditions for greater specialisation, technical change and spatial concentration of non-agricultural production. This process, which one would expect to be rather more pronounced in manufactures, will be further accentuated with improvements in transport (which reduce transport costs). The net effect might well be to increase dependence on non-local sources for meeting demand for non-agricultural goods and services by encouraging a shift in preferences favouring new products from outsides and/or extra local supplies for traditional products. On the other hand; countering these tendencies to displace local non-agricultural activists is the fact that the volume of trade and trade-related activities increases greatly as a result of commercialisation. It is difficult to hazard and *a priori* prediction of the balance of these influences on the overall level of non-agricultural employment. We will explore them empirically.

Before doing so, we must note the possibility that greater specialisation, and the differentiation of tasks which it implies, may affect the way people perceive and report their activity status. For instance, a carpenter or a cabbler who combines cultivation with his craft may not be able to specify the sector of his employment in terms of the categories used by those conducting the census or the sample surveys. To the extent such multiple occupation status is more common in traditional, relatively less commercialised communities, there is a problem of comparing industrial distribution of employment across communities/regions at different levels of commercialisation. This is especially true of the 'usual status' classification of workers; the problem is considerably less serious, but not eliminated, if one ascertains activity-wise time disposition by the week or, better still, the day.

A similar non-comparability of data on the reported sectoral distribution of personal days of employment may arise in situations where the incidence of wage labour is significant but both its incidence and forms vary. For instance a permanent farm servant does a wide variety of tasks. Some of it is clearly non-agricultural but in the usual survey they, and their labour time, probably tend to be recorded mostly under agriculture. A casual labourer on the other hand is more likely to differentiate between tasks he does if we are considering time disposition. In other words, it is possible that an area with a high level of casual wage labour may report a higher proportion of mandays in non-agricultural activity than another region with comparable incidence of wage labour as a whole but a relatively high incidence of permanent or attached farm servants.

Non-Agricultural Activity as a Residual Sector

One other possibility to be considered is that rural workers who cannot get adequate work in agriculture spill over into rural non-agricultural activities so that the latter acts—on the analogy of the 'informal sector' in urban areas—as a sponge for the excess labour. Such spill over may not manifest itself in traditional rural communities where family community and patron-client relations are strong. It seems more likely in situation where commercialisation has advanced and the wage labour system has become widespread inasmuch as both tend to weaken the traditional social mechanisms for taking care of the unemployed/ under-employed and increases the pressure on those who cannot find work in agriculture to seek other avenues of employment. And, in terms of the conventional supply-demand analysis of the labour market, if the absorptive capacity of agriculture and of urban areas is limited, the pressure of excess labour supply in rural areas will fall more heavily on rural non-agricultural sector which means that the level of the rural non-agricultural wage rate relative to the agricultural wage should be quite sensitive to the extent of imbalance between labour supply and demand in the rural areas. This imbalance is not easy to measure, but one may perhaps take the rate of rural un-employment, the higher is likely to be the share of non-agricultural sector in total rural employment and the lower the non-agricultural wage relative to that in agriculture. To recapitulate, we have identified the following factors which seem likely to influence the share of non-agricultural sectors in total rural employment:

1. Per capita rural demand for non-agricultural products. This is expected to be a rising function of (a) per capita income of agricultural classes and the degree of inequality in its distribution (b) the magnitude (and content) of public expenditure in rural areas; and (c) the degree of commercialisation (measured by the proportion of rural produce which is sold in the market). Commercialisation itself is likely to be an increasing function of (a), but depending in addition of the nature of technology used and on the importance of commercial crops.
2. Extra local demand for rural manufactures and services.
3. Rural demand for non-agricultural goods and services produced locally relative to the total rural non-agricultural demand. This seems likely to be inversely related to the degree of commercialisation (and perhaps also per capita income and its distribution) inasmuch as the latter affects preferences of consumers and induces changes in location, scale and technique of production. This means that the mean income of agricultural, population, inequality in income distribution and commercialisation have both positive and negative effects on aggregate rural demand for non-agricultural products. The net effect on is not easy to predict.
4. The extent of excess in labour supply relative to the requirement of workers in agriculture and exogenously determined non-agricultural activity.

In attempting to assess the strength of these influences in explaining the observed variations in rural non-agricultural employment, we have to contend with the problem of inadequate data.

Data and Their Limitations

We have estimates of rural non-agricultural employment from two sources, namely the Population Censuses and the NSS. The latter gives 3 different estimates (based on the concept of usual status, current weekly status and current daily status). While they differ from each other in absolute terms, the relative position of different states/regions is substantially the same by all measures. On the whole, the NSS estimates being based on more detailed and precise definitions of relevant categories are more reliable. But published tabulations are available only upto the state level thereby precluding analysis of variations at a more disaggregated level. Second, among the explanatory factors, no information is available on the magnitude of extra-local demand for rural manufactures/services even at the national level, let alone regional levels. The same is true of public spending in rural areas.

Third, we have state level estimates of net income from agriculture and allied activities and also of the value of crop production. The estimates for non-crop activities are at best notional. Those relating to crops are derived from measurement of actual output which, though by no means complete or accurate, are far better grounded than other elements. As per distribution the NSS gives estimates of inequalities in *consumption* for the rural population as a whole, but not separately for agricultural classes. Estimates of inequality in asset ownership and landholdings are also available, again for the entire population. The order of inequality in landholdings among those who have some land can be used for capturing the relative inequalities of income among agricultural classes—but clearly this is not entirely satisfactory.

Fourthly, the only available measure of commercialisation we have is the extent of monetisation in consumption. What we need is a measure of the quantum of farm produce sold as a proportion of the total farm production. Even if one were to take the former as a reasonable proxy for the latter, the latest year for which estimates are available (1963-64) is nearly two decades old and may not give a reliable picture of the situation in the seventies. The extent of commercialisation has therefore to be captured indirectly. It seems plausible to suppose commercialisation to be an increasing function of the level of output per head of agricultural population, the inequality in land distribution and the percentage of area under non-foodgrains. Though not quite satisfactory—it leaves out the effect of differences in the spread of modern agricultural technique which is intensive in the use of manufactured inputs—this is the best that can be done at present.

Finally since the requisite data for various explanatory variables are available only upto the state level (and the major states number barely 16), there is a limit to the number of explanatory variables which can be accommodated in the analysis.

Keeping all these in view, we choose the following as the candidate explanatory variables.

1. Crop output for head of agricultural population (CO/AP).
2. The Gini Index of concentration of operational holdings (LR).
3. The proportion of cropped area devoted to non-foodgrain crops (per cent NFG).
4. The unemployment rate in rural areas (PDUR). Their relation to the incidence of non-agricultural employment is examined by estimating a linear multiple regression.

The assumption is this CO/AP captures reasonably well the relative position of states in terms of the prosperity of agricultural classes (including both cultivators and labourers) as a whole, and that LR of land-holding captures thc differences in inequality in distribution of income. Both CO/AP and LR have also a bearing on commercialisation. We have distinguished between the incidence of non-agricultural employment among males and in the total workforce, as well as between workers classified by their usual and current status. In the case of unemployment, the current daily status estimates have been used.

The relevant data are summarised in Tables 10.13 to 10.14. The former relates to area 1971: The incidence of non-agricultural employment is taken from the 1971 census of population and the NSS estimate for 1972-73; the crop output estimates relate to 1969-70, the per capita figures being derived on the basis of the distribution of male workers between agriculture and other activities estimated from the 1971 population census; the inequality of landholdings and percentage of area under NFG relate to 1971. In Table 10.2, percentage of non-agricultural employment, unemployment rates and share of non-foodgrain crops relate to 1977-78, crop output estimates represent the average for 1975-76 to 1979-80; the LR relates to 1971-72. That the data for all variables in either tables do not related to the same point of time and that the data are subject to error argues for caution in interpreting the results.

TABLE 10.13 : Rural Non-Agricultural Employment and Related Factors by State 1971.

State	*Non-Agricultural Workers as Per cent of All Workers (NSS)*[1]		*Non-Agricultural Workers as Per cent of All Workers (Census)*[2]		*Crop Output Per Agricultural Population*[3]	*Moneti-sation of Consump-tion*[4]	*LR of Land Holdings*[5]	*Per cent Area under Non-Foodgrains*[6]	*Unemploy-ment Rate*[7]
	Total	*Male*	*Total*	*Male*					
Andhra Pradesh	22.1	23.5	20.9	24.4	411	62	.741	35.5	8.4
Assam	18.0	18.5	27.3	23.4	462	55	.583	28.5	1.7
Bihar	19.7	20.2	11.4	12.3	221	52	.644	10.1	8.9
Gujarat	19.3	21.4	16.2	18.2	353	65	.682	50.3	5.4
Karnataka	19.4	20.5	20.2	20.5	409	62	.679	33.3	8.1
Kerala	50.1	50.3	45.4	46.6	556	81	.678	68.9	24.5
Madhya Pradesh	10.7	12.5	11.0	12.8	360	45	.595	19.4	2.8
Maharashtra	32.4	34.8	14.3	18.2	308	66	.666	30.3	9.1
Orissa	23.8	22.4	16.9	16.2	418	52	.627	14.7	7.7
Punjab and Haryana	23.8	20.3	22.2	21.8	729	52	.754	27.0	5.0
Rajasthan	17.0	21.7	15.4	16.3	297	50	.558	24.0	4.1
Tamil Nadu	22.0	24.9	21.6	23.4	317	73	.719	33.7	10.8
Uttar Pradesh	19.7	20.4	12.9	13.5	370	46	.614	15.2	3.4
West Bengal	28.8	25.9	21.9	21.2	354	59	.643	12.6	9.9

Sources :

1. Relates to 1972-73, *Sarvekshana,* January 1979.
2. Census of India 1971.
3. Gross crop output 1969-70 (CSO est) + population 1971 + of all male workers engaged in agriculture 1971 CMIE (1975).
4. Relates of 1963-64. Estimates of NSS cited in Sankarnarayanan (1980).
5. Estimates from NSS, relates to 1971-72—excludes O class.
6. CSO Statistical Abstract.
7. Person-day unemployment rate for rural male, NSS, 1972-73.

TABLE 10.14 : Proportion of Rural Workers in Non-Agricultural Activities and Related Variables, States 1977-78.

	Percent of Non-Agricultural Workers				*Crop Output/Head Agricultural Population*	*Percent Area under Non-Foodgrains*	*LR of Land Holdings*	*Unemployment Rate*	
	Usual States		*Current Daily Status*					*Total*	*Male*
	Total	*Male*	*Total*	Male					
	(1)	(2)	(3)	(4)	(5)	(6)	(7)	(8)	(9)
Andhra Pradesh	17.3	19.7	23.2	24.1	444	30	0.741	11.7	9.0
Assam	14.5	14.5	15.5	15.1	407	23	0.583	1.6	1.6
Bihar	15.9	17.0	18.8	19.3	236	7	0.644	8.6	8.3
Gujarat	12.4	15.7	14.7	17.5	565	56	0.682	6.4	6.5
Haryana	22.6	22.9	22.3	23.5	1141	23	0.682	6.7	7.4
Jammu and Kashmir	22.7	26.0	25.8	28.7	362	10	0.436	6.8	6.6
Karnataka	15.6	16.9	17.7	18.6	493	37	0.679	9.8	8.3
Kerala	43.8	43.4	48.7	47.8	853	70	0.678	34.8	33.4
Madhya Pradesh	8.7	10.8	12.6	13.8	374	19	0.595	2.8	2.5
Maharashtra	14.7	19.6	19.7	23.9	433	26	0.666	7.8	6.2
Orissa	15.3	15.5	20.7	19.6	427	16	0.627	8.7	8.1
Punjab	23.1	23.2	21.7	23.0	1516	29	0.786	5.2	5.9
Rajasthan	13.4	18.0	14.6	19.7	371	29	0.558	2.8	3.2
Tamil Nadu	23.5	26.0	27.6	30.1	579	27	0.719	18.6	17.6
Uttar Pradesh	20.2	29.2	19.0	20.7	423	18	0.614	3.9	4.2
West Bengal	25.8	22.6	28.6	25.2	403	16	0.643	10.3	10.3

Sources :

1. Data from 32nd round of NSS related to 1977-78. Ratio of unemployed to working population according to current daily status classification, *Sarvekshana*, July-October, 1982, and January 1984.
2. Value of crop output per capita (average 1975-76 to 1979-80) estimated by Centre for Monitoring the Indian Economy (1985)+ per cent of all male workers employed in agriculture according to the 1971 census (1981 census figures on workforce and its sectoral distribution is as yet incomplete). The estimate of crop output covers all cereals gram, groundnut, rapeseed and mustard, cotton, jute, tea, coffee, potato, bananas and sugarcane.
3. CSO, Statistical Abstract, 1980.
4. Relates to 1071-72, Operational holdings excluding 'O' class.

1. Overall the selected explanatory variables account for 80 per cent of the variation in non-agricultural employment. This is true for males as well as for all workers, and whether one takes the usual status of workers or their current status.
2. In all cases, except current status of workers in 1971, the incidence of non-agricultural employment is seen to have a significant positive relation with crop output per head of agricultural population: that is, to say states with relatively prosperous agricultural population tend to have a high proportion of rural workers in non-agricultural activity.
3. However, other things being given, higher inequality of operational holdings seem to go with lower incidence of non-agricultural employment: This relationship is statistically significant in all cases except the current status measure in 1971.
4. On the basis of NSS data, the percentage of area under non-foodgrain crops has no significant association with non-agricultural employment. But the two variables show a significantly positive relation when we use the 1971 census estimate of rural non-agricultural employment.
5. In all cases, there is a significant positive relation between the rural unemployment rate and the incidence of non-agricultural employment and this association is much stronger than in the case of all other explanatory variables.

Given that we do not have a direct measure of commercialisation, and that the gross crop output is an incomplete and crude measure of net incomes of agricultural classes, caution is warranted in drawing inferences about the effect of the prosperity of the agricultural classes and the degree of commercialisation on rural non-agricultural employment. Refinements in this respect are contingent on more and better information becoming available. But there can be no doubt about the strong positive association between unemployment rate and non-agricultural employment in rural India. This would seem to give credibility to the notion that non-agricultural activities in rural areas may be acting to some extent as a residual sector absorbing labour which cannot find work in agriculture. If this were so, as we have argued earlier, it must be reflected in systematic variation of non-agricultural wage rates relate to agricultural wage with reference to unemployment. The relevant data on this are summarised in Table 10.16.

A plotting of the data shows no striking relation between the unemployment rate on the one hand and the non-agricultural wage. The non-agricultural wage rate is in general higher than agricultural wages in both years for which data are available, the difference being more prominent in 1977-78 than 1974-75. However, this may be more a reflection of differences in the coverage of the two estimates. The 1974-75 data relate to overall average daily earning of all made rural

TABLE 10.15 : Estimated Coefficients of Linear Multiple Regression of the Ratio of Non-Agricultural to Total Employment in Rural Areas and Selected Variables

		CO/Head	*LR*	*Per cent NFG*	*PDUR*	R^{-2}
1971 NSS-CDS	M	.0045 (.423)	–23.125 (.785)	.107 (1.073)	1.332 (4.891)	.784
	T	.0183 (1.67)	–18.573 (.774)	–.0092 (.09)	1.513 (5.431)	.789
1977 NSS-CDS	M	.0061 (2.241)	–26.51 (-2.294)	–.0489 (.799)	1.050 (8.423)	.871
	T	.0082 (3.272)	–28.482 (–2.681)	–.1065 (1.900)	1.138 (10.502)	.906
1971 Census US	M	.0240 (3.122)	–23.048 (1.369)	.2044 (2.860)	.8879 (4.546)	.876
	T	.0348 (3.396)	–42.416 (1.892)	.1668 (1.752)	.8983 (3.453)	.792
1977 NSS US	M	.0092 (3.525)	–31.705 (2.869)	–.0323 (–.552)	.9059 (7.596)	.862
	T	.0126 (3.9028)	–35.089 (2.563)	–.091 (1.263)	.9681 (6.931)	.827

Note : Figures in bracket are T values.
CDS = Current daily status.
US = Usual status.

wage labourers in agricultural and non-agricultural work; the 1977-78 estimates relate to average earnings of casual male labourers in manual agricultural operation other than ploughing, sowing, transplanting, weeding and harvesting, and to manual non-agricultural work. It happens that the farmer is in general lower than the average earnings for labour in 'other' agricultural work.

That apart, there seems at first sight no indication of the ratio of

TABLE 10.16 : Unemployment and Wage Rates in Rural India

	1974-75			1977-78[3]		
	PUDR[1]	*Average Daily Earnings (in Rs.) of Males*[2]		*PDUR*	*Average Daily Earnings (in Rs.) of Males*	
		Agriculture	*Non-Agriculture*		*Agriculture*	*Non-Agriculture*
Andhra Pradesh	8.4	2.66	2.80	9.0	3.50	4.39
Assam	1.7	4.02	3.29	1.6	5.12	5.42
Bihar	8.9	3.24	3.38	8.3	3.39	4.53
Gujarat	5.4	3.22	3.22	6.5	3.89	5.70
Haryana	*	*	*	7.4	5.75	6.06
Karnataka	8.1	2.87	3.16	8.3	3.04	4.15
Kerala	24.5	5.97	5.48	33.4	6.65	6.72
Madhya Pradesh	2.8	2.42	2.47	2.5	2.76	6.79
Maharashtra	9.1	2.64	2.68	6.2	3.15	4.53
Orissa	7.7	2.64	2.50	8.1	3.07	3.75
Punjab	5.0**	5.60**	5.10**	5.9	7.08	7.71
Rajasthan	4.1	3.85	3.27	3.2	4.33	5.37
Tamil Nadu	10.8	3.69	3.56	17.6	3.89	4.32
Uttar Pradesh	3.4	3.21	3.66	4.2	3.63	5.20
West Bengal	9.9	3.49	3.27	10.3	3.98	5.37

Notes:

1. Relates to 1972-73. Estimates from *Sarvekshana.*
2. Relations to all agricultural and non-agricultural operations respectively. Estimates from report of the Rural Labour Enquiry.
3. NSS 32nd round estimates, *Sarvakshana,* April-June 1981. Average daily earnings relate to manual work in other agricultural operations' and in manual non-agricultural work.

* Included under Punjab.

** Includes Haryana.

non-agricultural to agricultural wage being inversely related to unemployment rate, this relation being implicit in the residual sector hypothesis. However, there is indication of a more complex pattern. Assam, Punjab and Haryana, with relatively low unemployment and high wage rates, also show very low differences between non-agricultural and agricultural wage rate. Among the remainder, the scatter suggests that in states with relatively high unemployment rates (8^+ per cent in 1974-75 and 6^+ per cent in 1977-78) there is a noticeable tendency for the ratio of non-agricultural to agricultural wage rates to narrow as unemployment increases.

Changes Over Time

In the post-independence period, several of the factors which we have identified as having a bearing on rural non-agricultural employment have shown major systematic changes. In the first place, level of crop output relative to agricultural population has increased progressively if slowly; output of non-foodgrains crops has grown nearly as fast as total output; dependence on purchased, non-rural inputs relative to output has increased rapidly from less than 5 per cent in 1960-61 for the country as a whole to 15 per cent in 1980-81; and the extent and quality of transport network linking rural with urban areas has improved. All of this points of increased commercialisation of the agricultural sector. Throughout this period, the proportion of wage labour in the rural workforce has risen and the wage labour is getting increasingly casualised.

Growing commercialisation of agriculture must have intensified the process of widening and integration of rural markets for manufactures leading to reorganisation and relocation of traditional rural industry and stimulating technological change. It must also have made it easier for newer products from urban areas to penetrate the rural market and bring about changes in the consumption patterns of the rural areas. That such changes have taken place is apparent to even casual observation though their magnitude and impact remains to be assessed.

The impact of relatively fast growth of rural commerce transport and manufactures on rural non-agricultural employment have been dampened by technological changes which are taking place, partly autonomously and partly as a consequence of the widening of markets. The rapid growth of power-looms over handlooms; of mechanised over traditional techniques of processing farm produce, and rapid spread of mechanised forms of transport are cases in point. These changes have second round consequences tending to depress the growth of activity and employment in supporting activities.

On the other hand, the rapid expansion of public sector activity in rural areas has led to a steep rise in the number of people employed in rural areas directly by the government in administration and developmental activities. Moreover, expenditures on various development programmes (including anti-poverty programmes) generates incomes among their participants (including government functionaries and local leaders and their agents). The rapid expansion in public expenditure in rural areas is therefore a new and growing source of rural incomes which in turn generates successive rounds of increase in the demand for several non-agricultural products and services at least part of which are made locally. We do not however have the information necessary to estimate the extent of the expansion of these accounts, or its impact by region or even by state.

The extension and improvement of road network and transport facilities between rural and urban areas may have brought about some changes in the pattern of rural-urban migration. Better roads and modern means of transport reduce the time required for travel to and from nearby areas and also reduce the cost of such travel thus making it easier for workers to continue residing in their village homes and commute to their jobs in the nearby towns. The incentive for doing so would be stronger if the costs involved in the worker shifting residence to urban areas have risen. Again, the absence of information, this proposition is necessarily rather speculative.

That, along with these shifts, considerable changes in technology and organisation have taken place in rural industry and transport is apparent enough. The data from the decennial censuses suggest that the relative importance of household industry in rural industrial activity (proportion of workers in this sector) have declined substantially between 1961 and 1981 even as rural non-household industry has grown in absolute are relative terms. But it is impossible with the sketchy tabulations of the 1981 census currently available to say much about the specific segments of industry and regions where the shifts have occurred (Eapen and Vaidyanathan, 1984).

The currently available information being inadequate even to get a reasonably definite picture of the trend in rural non-agricultural employment in different parts of India, it is even more problematic to relate them to changes in various factors discussed above. The decennial censuses of population show a fall in proportion of rural workforce engaged in non-agricultural activities between 1961 and 1971, followed by a rise in 1981 but to a level below that of 1961. There are however well known problems of comparability of data between census which make them not quite dependable as the basis for judging trends. The NSS of the other hand has taken considerable care to ensure comparability of concepts and definitions especially since the early 1970s and is probably much more reliable. It points to substantial and sustained rise in the share of non-agricultural activities in rural employment during the seventies both at the national level and in a majority of the states. Seven out of the 16 states reported a decline in non-agricultural employment relative to total employment between 1972-73 and 1977-78, but all of them show a rise in 1983; in several cases the rise more than compensates for the earlier decline (Table 10.17).

TABLE 10.17 :Changes in Non-Agricultural Employment and Unemployment in Percentage of Rural Workforce (Male) 1972-73, 1977-78, and 1983-84.

	1972-73 to 1977-78 Change in			*1977-78 to 1983 Change in*		
	Non-Agricultural Employment[1] (Per Cent Points)	*Foodgrain Production (Per Cent)*	*Unemployment Rate[1] (Per Cent Points)*	*Non-Agricultural Employment[2] (Per Cent Points)*	*Foodgrain Production (Per Cent)*	*Unemployment Rate[2] (Per cent Points)*
	1	2	3	4	5	6
Andhra Pradesh	+ 1.8	+ 34	+ .3	+ 4.8	+ 28	+ .3
Assam	− 9.1	+ 2	− .2	+ 6.3	+ 11	+ 1.9
Bihar	− 1.2	+ 6	−1.4	+ 2.8	− 2	− .6
Gujarat	− 3.1	+ 75	+ 1.0	+ 4.5	+ 50	− 1.1
Haryana	+ 5.4	+ 31	+ 3.5	+ 5.1	+ 29	+ .4
Jammu and Kashmir	+ 7.2	+ 11	− 2.7	+10.2	+ 1	+ 9.8
Karnataka	− 1.3	+ 58	+ 0.3	+ .8	+ 1	− 1.8
Kerala	− .8	− 6	+ 2.6	+ 1.7	− 4	− 3.6
Madhya Pradesh	+ .9	+ 16	− .3	+ 2	+ 24	+ .1
Maharashtra	−10.8	+243	− 2.2	+ 2.8	+ 5	+ .3
Orissa	− 1.8	+ 11	+ .2	+ 7.1	+ 23	+ .8
Punjab	+ 5.3	+ 35	− .0	+ 5.3	+ 45	+ 1.0
Rajasthan	+ .6	+ 39	− .6	+ 4.2	+ 41	+ 1.6
Tamil Nadu	+ 6.6	+ 8	+ 6.7	+ 7.6	− 20	+ 5.2
Uttar Pradesh	+ 1.1	+ 17	+ 1.0	+ 2.3	+ 38	− .4
West Bengal	+ .8	+ 32	− .2	+ 5.3	+ 4	+ 7.1

Note :

1. The correlation coefficients between change in non-agricultural employment on the one hand and the changes in foodgrain output and unemployment on the other are as follow :

	1972-77	1977-83
Foodgrains output	−.55	−.11
Unemployment	.38	.87

Sources :

1. Relates to population in the age group 15-59 years. 1972-73 estimates are from the 28th round of NSS. 1977-78 estimates are from the 32nd round of NSS. *Sarvekshana.*
2. Relate to population aged 5 years or more estimates for 1977-78 from *Sarvekshana*, January-April 1981 and for 1983, NSS Report No. 315.

Between 1972-73 and 1977-78 practically all states recorded a rise in agricultural production, and yet several recorded a decline in the ratio of non-agricultural to total employment. With the significant exception of Maharashtra and Gujarat, the states in the latter category recorded relatively marginal rise in output. There is no association between the direction of change in unemployment rate and in non-agricultural employment during this period.

Between 1977-78 and 1983 on the other hand, practically all states report a rise in the non-agricultural employment ratio, including those which have experienced little or no increase in foodgrain production. Again there is no strong association between growth of foodgrain production and changes in the ratio of non-agricultural to agricultural employment. The latter however seems to be positively correlated with change in rural unemployment—the correlation coefficient being high (.87) and significant. That is an increase in unemployment has generally gone with a rise in the ratio of non-agricultural employment.

IV. CONCLUSIONS

While the above findings seem to offer some support to the residual sector hypothesis, the question remains why the non-agricultural sector fails to absorb all the available labour in excess of agriculture's needs. Could it be that the observed pattern of association between unemployment and non-agricultural employment reflects structural changes which go with commercialisation and the growth of wage labour force? There seems to be some basis to such an interpretation in as much as the unemployment rates happen to be positively and significantly correlated with the incidence of wage labour and with the degree of commercialisation : the latter two factors are themselves positively correlated.[16] If forces resulting in greater commercialisation lead to an increase in the incidence of wage labour and casualisation of wage labour, the concomitant break-down of traditional mechanism to take care of the 'surplus labour' leads to latent unemployment coming more and more into the open. The very same process could, as explained earlier, lead to a greater specialisation and differentiation of tasks performed within the rural labour force resulting in a rise in the reported distribution of employment in non-agricultural activities relative to agriculture. Which of those interpretations in nearer the truth cannot be established without much more disaggregated data on the pattern of regional variation, supplemented by detailed micro studies of the pattern of employment, structure of the labour market and the way they change in different types of situations.

Note : Reproduced from *Economic and Political Weekly,* Vol. XXI, No. 52, Review of Agriculture, December 27, 1986.

NOTES

1. This category is called 'main workers'. Including 'marginal workers' raises the number to 229 million, the difference being mostly under females.
2. Main plus marginal.
3. This estimate of employment in persondays and its composition is from the NSS and relates to main workers.
4. *Sarvekshana,* J-O 1981: S37-S38.
5. In terms of 'usual' status, about 90 per cent of the persons (main and marginal works) who report their usual activity to be agriculture and related activities, report themselves to be in agricultural production (including plantation) a little over 9 per cent in livestock and the rest in fisheries, forests, etc. I am greatful to K. Sundaram for giving these figure which he reconstructed from unpublished tabulations.
6. Sundaram has drawn my attention to the possibility that part of the decline in CDS employment rate may be due to the introduction in 1977-78 of a separate and explicit category (93) for those combining domestic work with free collection of fuel wood, etc. In 1972-73 since this category was not recognised, some of the fuel wood collection, etc, may have been recorded as 'work'. However, this would make a significant difference only to the estimate for females.
7. Note that the estimates relate to different periods for different districts. In general they represent 3 year averages. These and other questions concerning concepts and coverage of data are discussed in Vaidyanathan (1978).
8. In the case of state cross-section, since relevant information was not accessible, we have used foodgrain yield/ha.
9. The proportion of area devoted to paddy and sugarcane has been taken as the index in this study. Both are known to be far more labour intensive than most other crops. One could easily identify several other high labour intensity crops (tobacco, vegetables, chillies, onions, etc) but their importance in terms of percentage of area allocated is likely to be negligible at the state level.
10. For a comprehensive review of the literature see Rakesh Basant (1985).
11. The family labour based farms interested in maximising total income carries labour use much further than a farmer depending on wage labour would, because the latter is interested in maximising his net income after meeting wage costs.

12. Available estimates for 1983 are incomplete in that they relate only to 2 sub-rounds. Also, as already mentioned the NSS estimate or rural population growth between 1977 and 1983 seems out of line with census data.
13. On the basis of NSS data for 1977-78 about two-fifths of rural non-agricultural employment is in mining and manufacturing (the former being insignificant); approximately a quarter each in (a) trade, transport and communication, and (b) community, social and personal services. About 8 per cent is reported to be in construction.
14. Sankaranarayanan (1980) in an unpublished paper drew attention to the existence of these variations and went on to consider how far these are associated with the prosperity of the rural population and commercialisation of rural economy (measured by monetisation of consumption). While he found the former to be relatively non-significant, he noted a positive and often significant, association between commercialisation, and the share of non-agricultural activities (both overall and by major subcategory) in rural employment.
15. A rough estimate based on the census workforce data shows that the number of rural male workers engaged in electricity, communication, public administration and defence, education and health has increased from 2.5 million (2.4 per cent of all rural male workers) in 1961 to 5.3 million (3.9 per cent of male workforce) in 1981.
16. From Table 25.13, the correlation between unemployment rate and incidence of wage labour is calculated at 0.71; the coefficient of correlation between unemployment and degree of monetisation consumption is 0.81; and between the latter and the incidence of wage labour 0.71. Some caution is needed in interpreting these correlations because monetisation of consumption does not give a complete picture of commercialisation and also because it is based on rather old data.

REFERENCES

Agarwal, Bina, (1980), 'Tractorisation, Productivity and Employment : A Reassessment', *Journal of Development Studies,* 16, 3.

Alagh, Y. K., *et al,* (1978), 'Agricultural Growth and Manpower Absorption' in Bardhan *et al*, 1978.

Bardhan, P. K., (1978), 'On Labour Absorption in Asian Rice Cultivation...' in Bardhan *et al*, 1978.

Bardhan, P. K., *et al,* (1978) "Labour Absorption in Indian Agriculture : Some Explanatory Investigations" (ARTEP, ILO, Bangkok).

Basant, Rakesh, (1985) "Agricultural Technology and Employment in India : A Survey of Recent Research" (Gujarat Institute of Area Planning, Ahmedabad, Working Paper).

Binswanger, H. W., (1978), "The Economics of Tractors in South Asia : An Analytical Review" (ICRISAT, Hyderabad).

Bartsch, W. H., (1977), "Employment and Technological Choice in Asian Agriculture" (ILO, Geneva).

Eapen, M. and Vaidynathan A., (1984), "Structure of Employment in Indian Industry; Some findings from Census Data" (CDS, Trivandrum, Working Paper 199)

Ishikawa, S, (1967), "Economic Development in Asian Perspective" (Kinokiniya, Tokyo). 1978 : "Labour Absorption in Asian Agriculture" (ARTEP, ILO, Bangkok).

Jose, A. V., (1982), "Growth and Fluctuation in Indian Agriculture 1952-53 to 1978-79" mimeo, unpublished).

Mehra, S., (1976), 'Some Aspects of Labour Use in Indian Agriculture', *IJAE*, xxxi-4.

Sen, A. K.,. (1962), 'An Aspect of Indian Agriculture', *Economic Weekly.*

Sinha, J. N., (1979), 'Employment and Agriculture in Draft Plan 1978-83', *EPW.*

Srinivasan, T. N., (1971), 'Farm Size and Productivity : Implications of Choice under Uncertainty', *Sankhya,* Series B, Part 4.

Sankaranarayanan, V., (1980), "Inter-State Variations in Rural Non-agricultural Employment : Some Tentative Results" (CDS, Trivandrum, Working Paper 104).

Vaidyanathan, A., (1978), 'Labour Use in Indian Agriculture : An Analysis Based on Farm Management Survey Data' in Bardhan *et al*, 1978.

Vaidyanathan, A., (1985), "Water Control Institutions and Agriculture : A Comparative Perspective", *Indian Economic Review.*

INDEX